T0347134

For Velvet, and all of our children

DAMON GAMEAU

Pan Macmillan Australia

CONTENTS

FOREWORD
BY PAUL HAWKEN

PROJECT DRAWDOWN

AUTHOR, *DRAWDOWN: THE MOST COMPREHENSIVE PLAN EVER PROPOSED TO REVERSE GLOBAL WARMING*

The physical laws that inform the science of global warming have long been known. In 1856, Eunice Foote demonstrated that carbon dioxide was the greenhouse gas that would retain the most heat and create a warmer climate if its emission increased. It was a far-sighted scientific discovery. Since then, we have placed 970,687,671.8 tonnes of CO2 into the atmosphere, an enormous planetary gamble. We were thoroughly warned by climate scientists about the danger and the results are plain to see: intense heat, rising oceans, melting ice, formidable droughts, biblical flooding, conflicts, migrations, food shortages . . . the list goes on. These outcomes we know, see and hear about almost daily. And this is just the beginning. We face a problem so immense that we are staggered by its enormity and understandably feel powerless to prevent or address it. The question Damon Gameau asks is simple: Why look only at the problem?

Every problem is a solution in disguise. No exception. Given that global warming is the most gnarly, super-wicked problem ever faced by civilisation, it stands to reason that it contains an extraordinary number of solutions. The narrative around global warming has been a litany of probabilities of what is wrong, and what will go wrong – when, where and how. That is good science. What *2040* explores are the extraordinary solutions. They are invitations to a stunningly transformed world – a brilliant array of possibilities that are hidden and contained within the problems. That is good science too.

2040 is a vista into a remarkable future wherein imaginative and practical solutions to global warming are not penance but promise, not obligations but

'Every problem is a
solution in disguise.'

opportunities, not inhibition but innovation. I cannot imagine a more precious gift to our children than a view of the future that challenges, inspires and delights. The modern world often acts and thinks as though it is paradigmatically stuck, that this is the best of all possible worlds, and that the metes and bounds of practical change lie within a narrow range of what is possible. There are many who believe that moving too far from business-as-usual means we lose much of what we want. They are right in that much will be 'lost' if we seriously address and reverse global warming. However, we will not lose what we truly value. What will disappear is the collapse of biodiversity, the devastation of our fisheries, the clear-cutting of forests, the poison-soaked lands of industrial agriculture, the dying rivers, the poverty of billions, the joblessness of mothers and fathers, the despair in our youth, the suffering of the landless and hungry.

What we see and read in *2040* is that the most viable path to reversing global warming is to address current human needs. Yes, we certainly do not want to exceed 2°C in global temperature rise, the target stated in the Paris Agreement, but that is an abstract goal. Human beings need, and respond to, solutions that improve their security, income, health and habitats. If you have come to think that addressing global warming is mostly about wind turbines and solar panels, you will find in Damon's work a way of seeing the future that is a promise, not only to his daughter, but to all daughters and sons, to all creatures and habitats, to all who seek meaning and purpose in a world that seems adrift and confused. When going into darkness, the first rule is to bring illumination. Take Damon's message as your flashlight to an extraordinary future.

Language is the means by which a civilisation passes its values to a new generation. It allows human beings to connect with each other, to identify and give meaning to objects, and to generate cultural norms and patterns of behaviour that promote cohesion and ensure survival.

For a huge part of our existence, humans have ascribed great meaning and significance to the natural world. Our Aboriginal ancestors referred to themselves as 'custodians of the land' and revered the great spirits carved into the landscape. The ancient Egyptians saw nature as a giving parent and the Native Americans referred to 'Mother Earth' and 'Father Sky', seeing themselves as part of a web of life. In ancient China, people spoke of being 'reverent guests' of the land. The Chinese explorer Admiral Zheng, of whom we hear far too little, sailed the world with 27,000 men on 300 enormous ships well before Columbus did. He didn't dominate those he came across, but wished to explore the marvels of the natural world, exchange goods and bring back new plants and animals, such as zebras, camels and giraffes, to his homeland.

Fast-forward to the scientific revolution of the early 17th century and a new understanding, language and meaning of the natural world emerges. Although the spread of Christianity had already begun the process, it was philosophers and scientists such as René Descartes and Francis Bacon who really helped shift the metaphor of Earth from a nurturing parent to an 'object' to conquer and rule. 'We must hound Nature in her wanderings,' said Bacon, 'storm and occupy her castles'. He spoke of putting Nature in 'constraints' and, perhaps most symbolic for our current dilemma, entering and penetrating her every corner and hole.

We have been blessed with the wonders of science since this time but a new way of seeing the world has also snuck in the back door. A way of seeing nature as a machine, as separate from us, and as something to be exploited.

Metaphors have enormous power to shape our reality and define our behaviour. Our survival may depend on us urgently reclaiming old metaphors.

The Patterning Instinct is a beautiful book by Jeremy Lent. He describes the significance of meaning this way: if viewed solely through a reductionist, scientific lens, the works of Shakespeare would simply be 26 letters arranged in various sequences on a page. What evokes the wonder, emotion and beauty is the *meaning* we give to those arrangements of letters and words.

Since the scientific revolution, our planet has been viewed through a similar lens – as a rock (with some plants and the odd animal on it for us to eat) drifting through a cold and brutal universe. But to save the planet, perhaps we need to see it once again as many of our ancestors did – as a complex, interwoven system of life, and as a home. It is not 'the' environment, it is 'our' environment. We are not separate from it and never have been.

MY MOTIVATIONS

If something doesn't have meaning, or significance, then why would you defend it? Why would you honour it? I think this sums up my own lack of engagement with most environmental issues over the last 30 years. I was always busy giving meaning to other things in life, such as wooing the opposite sex, earning money for life's essentials like rent, food, alcohol and cigarettes (the priority of this order varied widely), trying to forge a successful career, going on holidays or simply getting from A to B as quickly as possible. I did like being outdoors, away from the city, but nature was just 'there' (often a bit dull); it was a place to kick a football, or to stop and take a quick wee on a long drive.

As for 'climate change', well, that got shoved way down the list. It was far too big for me to contemplate or engage with (what could I do about it anyway?), and I didn't really think it would impact me in my lifetime. Plus, amongst the wide range of web content that a developing male gravitates to, many sites told me that it was either a hoax or some kind of global government conspiracy. Which, to be honest, suited me perfectly, as it justified even less engagement on my part.

I recently spoke to the environmental psychologist Renee Lertzman who gave me another perspective on my lack of willingness to engage. 'Many people are experiencing what I'd call a "latent" form of climate anxiety or dread,' she told me. 'They may not be talking about it much but they are feeling it. It's important to remember that inaction is rarely about a lack of concern or care, but is so much more complex. Namely, that we westerners are living in a society that is still deeply entrenched in the very practices we now know are damaging and destructive. This creates a very specific kind of situation – what psychologists call cognitive dissonance. Unless we know how to work with this dissonance, we will continue to come up against resistance, inaction and reactivity.'

This nailed it for me. My brain couldn't work out how to solve a problem in which I was so fundamentally implicated. I like flying in planes, driving my car is super convenient, drinking water from plastic bottles is often handy, and I do quite like Nikes to run in. Are you telling me these actions may be contributing to ecological collapse? No thanks, I'll be at the bar discussing girls and football.

But all that changed with a trip to the Amazon. (Which I'm sure is a sentence that has been uttered quite a few times throughout history.) It was a two-week stay in the heart of the jungle with my wife, Zoe. We had no technology and minimal western influence. It was just us, a giant tarantula, a billion other intimidating critters and a charismatic tour guide who had never left the jungle in his 21 years. This is where the concept of 'meaning' first struck me. To see such a density of nature, to understand how much life the jungle supports, and to feel an openness and wonder in its presence was a transformative experience (drinking deeply from a bowl of ayahuasca 'fruit punch' may also have been a factor). Most poignantly, I learnt of the jungle's reliance on a web of interconnectivity for survival. The trees 'talk to each other' via a network of fungi that grows between their root systems. Their foliage creates shade that keeps the ground moist. This moisture evaporates to form clouds, which increases rainfall and the cycle continues. Disturb just one of these processes and the system collapses. Every tree is valuable to the whole forest and that's why healthy trees will often send nutrients to sick trees until they recover.

Returning home from that trip, it was impossible not to notice how disconnected from nature I had become. Living in inner-city Melbourne, I did see the odd bin bird or slightly withered street tree, but my most regular views of 'nature' were pictures of tigers and elephants on the Melbourne Zoo billboard at the tram stop, or a glimpse of the Puma logo on my sneakers while tying my shoelaces. And, clearly, I'm not alone. A study from the Lawrence Berkeley National Laboratory found that US residents spend around 86 per cent of their time indoors (home, garage, vehicle, work, shopping centre, bar, restaurant, garage, home, then repeat . . .). What chance does nature have of us standing up for it if it no longer appears regularly in our consciousness? Out of sight, out of mind.

And so began my personal journey of reclamation. After the post-jungle boost to my entire wellbeing, I began to find other ways to immerse myself in nature to recapture that feeling. Ocean dunks helped. Moving closer to a forest helped, and so did switching off my phone for at least one day each weekend just to stare out a window or go for a walk (this was surprisingly tricky early on). As nature returned to my sight, it also returned to my mind and I gravitated to more books, articles and television shows about it.

And this, my friends, is where our story begins . . .

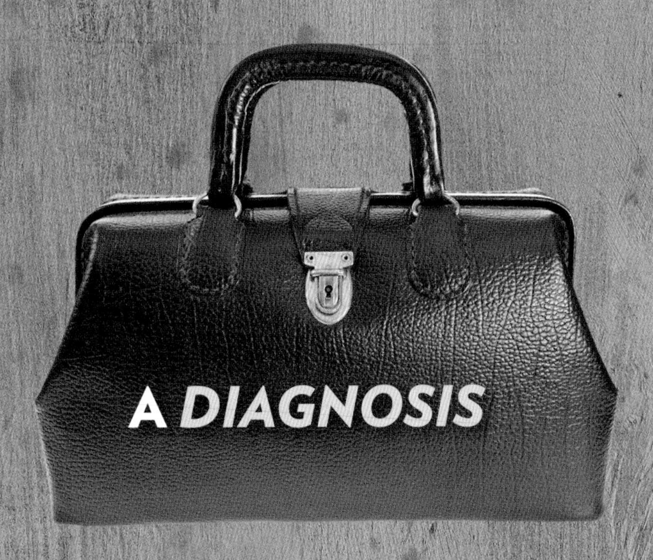

A DIAGNOSIS

Some of you will be aware that I spent a chunk of my life trying to understand why humans are becoming sicker and more overweight. In making *That Sugar Film*, I learnt that added sugar has left its sticky fingerprints all over the crime scene. As I was doing my research for *2040*, I began to see the similarities between the two projects. While we struggle to maintain our own health, the planet is also in dire need of a health intervention.

So let's imagine the planet has a doctor's appointment. It has just had a thorough examination and the grim-faced doctor has asked it to take a seat.

'Now, Earth, the news isn't great. Overall, you have a pretty bad fever. Since we saw you in 1910, your temperature has gone up 1°C. We measured this in your Australia region which, as you know, is 'down under' – so apologies again for where we stuck the thermometer. Unfortunately, you are on track to raise your temperature by at least another 1°C over the next few decades if you don't change your ways. This would see the devastation of habitats and the extinction of many of your precious living things.

'This fever is also contributing to other symptoms,' continues the doctor. 'Your oceans are not only warmer (over 90 per cent of your excess heat is being absorbed there), but also 30 per cent more acidic than when we first saw you a couple of hundred years ago. I believe this is when you began your 'Industrial Revolution diet'. These impacts are affecting your coral reefs, which, as you know, cover less than 1 per cent of your ocean floor but house almost a quarter of your marine life. (We'll get to plastics later.)

'Since this I-Rev diet, your delicate atmosphere has 40 per cent more carbon in it. This means it now has over 410 parts per million (and climbing rapidly) compared to the 280 parts per million it had when we first saw you pre I-Rev diet. This is a major concern. As the sun heats your surface, that heat is getting trapped by the excess carbon dioxide (CO2) and other greenhouse gases in your atmosphere. The excess heat not only leads to an increase in wildfires and more severe droughts, but the hotter air also holds more moisture, leading to more intense storms and heavier rainfall and snowfall.

'This heat also means your ice sheet in Greenland is losing around 300 billion tonnes of ice every year (that's more than the weight of one of your largest pimples, Mount Everest). When combined with your warmer oceans (which expand), melting ice is contributing to a sea level rise that will have a cataclysmic impact on the lives of hundreds of millions of your inhabitants. Kind of important.

'An additional related symptom is the health of your topsoils. These are critical to providing food for your inhabitants, and they are diminishing rapidly. I've consulted the other doctors and we estimate roughly 60 years of topsoils remain at your current levels of agricultural activity. And your nitrogen and phosphorous levels are off the charts. You have been using a lot of chemicals to offset the damage to your soils, and these chemicals are now appearing in your rivers and oceans.

'Another worrying symptom is shortness of breath, due to rabid deforestation. Your forests are effectively your lungs, yet you're losing a football field-sized plot of Amazon rainforest every 60 seconds (mainly for agricultural purposes, predominantly livestock). This forest removal also means you're losing biodiversity, with species of animals and plants becoming extinct at rates between 1000 and 10,000 times faster than natural extinction rates.

'So, Earth, after decades of debate and many heated arguments, my colleagues and I have recently reached a 97 per cent consensus on the major cause of your sickness.

'The first thing you absolutely *must* do is to lay off the smokestacks. I know upping your fossil-fuel intake served you well in the early stages of your I-Rev diet and you achieved some great things, but we didn't have any long-term studies of your diet available then and so couldn't foresee the trouble you now find yourself in. (To be fair, some of my colleagues did foresee a potential disaster, but they were ignored or silenced.)

'You're also growing a little too fast in terms of both population and economics for your own good. This means you are unable to replenish the resources you are using each year. You currently require 1.7 Earths' worth of resources in a single year due to the overfishing, mineral extraction, deforestation etc that is taking place. This is making it really tough for the majority of your living things.

'But it's predominantly the smokestacks that are driving your fever. The digging up and burning of stored carbon is pouring more than 40 billion tonnes into your atmosphere each year, so we need to stop more going in and also remove the excess carbon that's already there.

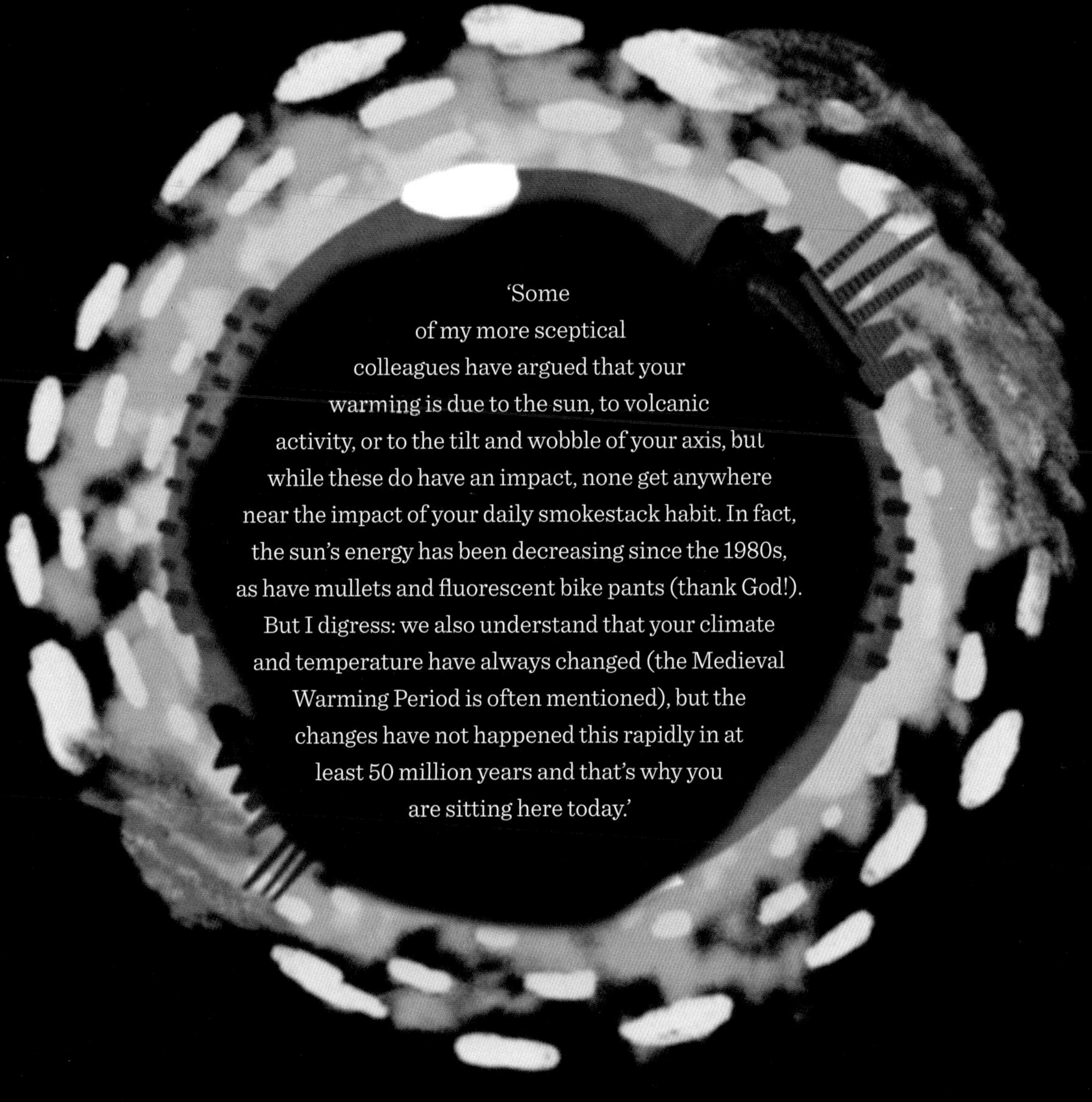

'Some
of my more sceptical
colleagues have argued that your
warming is due to the sun, to volcanic
activity, or to the tilt and wobble of your axis, but
while these do have an impact, none get anywhere
near the impact of your daily smokestack habit. In fact,
the sun's energy has been decreasing since the 1980s,
as have mullets and fluorescent bike pants (thank God!).
But I digress: we also understand that your climate
and temperature have always changed (the Medieval
Warming Period is often mentioned), but the
changes have not happened this rapidly in at
least 50 million years and that's why you
are sitting here today.'

'We'll lose more species of plants and animals between 2000 and 2065 than we've lost in the last 65 million years.'

PAUL WATSON, SEA SHEPHERD FOUNDER

A NEW STORY

Now imagine it was you in that doctor's office receiving the dire news, but that you were offered no treatment. You would leave the doctor feeling devastated by your diagnosis, but with no idea what you must do to get better.

This pretty much sums up the current media narrative around environmental issues. Whether it's in social media, the news or climate documentaries, the analyses of our predicament and future projections are almost entirely horrific. This diagnosis has its place, but as a motivational tool for kickstarting the public into action, again I turn to the environmental psychologist Renee Lertzman: 'Neuroscience and the mental health field is now saying that when information has a charge to it that brings up fear, guilt, anxiety or worry, even confusion, we cannot process the information very well. Cognitively we become impaired because the limbic system is activated and we lose access to our prefrontal cortex, which is where we problem-solve, are creative, make connections and can look ahead to the future.'

I am a dad and my four-year-old daughter's future matters to me deeply, so I decided to try and find an alternative to the doom and gloom stories we are currently bombarded with. I wanted to find the 'chicken soup and cold flannel' equivalents for our sick planetary home and show my daughter what her future could look like if we embrace those solutions today. I didn't want this to be an exercise in wishful thinking, where I tell her that everything will be fine. It's important that all of us acknowledge how scary some of the information we are hearing is and that it's okay to feel overwhelmed or upset. I wanted the solutions I found to be based in truth and to be realistic. The solutions had to already exist today in some form and they had to be scalable. As I say in the film, this was an exercise in 'fact-based dreaming'.

My mission took almost three years of research and planning, including travel to 14 countries and hundreds of interviews with scientists and other experts. And in case you immediately think, 'That must have used a truckload of carbon', we did offset our emissions with certified carbon credits, plus we planted a small native forest to sequester a further 90 tonnes of carbon by 2040.

The aim of this book is to provide you with genuine hope (as opposed to the hope that often gets wheeled out at election time). It is a book that will explain some of the things you can do right now to make an enormous difference to the future of the planet. It is a book that also reveals some of the amazing and inspiring things that people are already doing.

Make no mistake, healing our planet will be a complex operation. It will require a multidisciplinary approach from dedicated teams of surgeons at the grassroots level and from inside the halls of power. But this operation presents a never-before-seen opportunity to come together and change the course of history.

We know the diagnosis, so let's start telling a new story. A story about the solutions that can regenerate our planet.

'The ultimate test of
a moral society is the
kind of world it leaves
to its children.'
DIETRICH BONHOEFFER

THE VOICES OF 2040

I figured the first thing I should do when imagining a better future for our daughter is consult the generation that would be sharing it with her. I wanted their hopes and aspirations to help guide the arc of the story I would tell. My question to them went like this: 'So, when you grow up and you are roughly the same age as your mum or dad, what would you like to see in the world?'

I spoke to about 80 children between the ages of six and 11. They were predominantly from Brooklyn in New York, Oberlin in Ohio and Arusha in Tanzania, as well as Stockholm, London, Melbourne and Singapore. What I found both enlightening and alarming about these interviews was how articulate these children were around the issues concerning our planet. When I was their age, my biggest concern was getting to the final level of Donkey Kong or deciding when to eat the bubblegum nose on a Bubble O' Bill. As I listened to the children my thoughts jumped between 'They shouldn't be aware of this stuff!' to 'Look out when these kids start voting!'

Jonas and Jonathan Salk talk about our being at an 'intersection point' between two world views. On the one hand, there's the postwar generation's belief in growth as the key to prosperity. This is completely understandable when you look at the multiple benefits of this approach to individuals and society over the past 70 years. However, a new emerging viewpoint, often expressed by millennials, is that growth at all costs doesn't work on a planet with finite resources. This is no surprise when you consider that these young people have grown up with news stories of climate change, oceans polluted with plastic and resources becoming more scarce. I got a strong sense of this from the children's interviews. Their framing of the world and cultural norms are fundamentally shifting. This should give us all enormous hope for the future, as long as we don't drive the car off the cliff before they get control of the wheel.

After about a year of researching solutions, I was in Colorado for a gathering of environmentalists. (The collective noun for environmentalists could vary enormously depending on your political persuasion. I chose 'gathering'.) It was there that I was introduced to Paul Hawken. Paul is well known for many things, in particular his *New York Times* bestselling book *Blessed Unrest*.

Paul spoke on stage for an hour with such eloquence, compassion and intelligence that I knew he would have an enormous influence on my journey. He was the first person to clearly articulate for me what we need to do to begin healing our planet. Firstly, we must dramatically lower greenhouse gas emissions, and secondly, we must sequester or draw down the excess carbon that's already in our atmosphere by storing it in plants and soil.

Paul is the founder of a global climate initiative called Project Drawdown, which has created a map, measure and model of the top 100 solutions to reverse global warming. It was six years in the making, involving 200 experts from 22 countries, and every solution had to already exist and be scalable. You can understand why I felt so relieved sitting in the auditorium. On stage Paul explained that even without the threat to our climate, we would still want to implement 98 of the 100 solutions because of the cascading benefits to communities, children and women, animals, soil, food, air, water and jobs. Sold.

Project Drawdown completely reframes the narrative around global warming. It sets it up as an opportunity for change. It outlines clear solutions, highlighting their impact on reducing carbon while also analysing their cost-benefits. It is a remarkable achievement and a ready-made action plan for global transformation. I can report with some relief that it is already being discussed in some very high places.

I had a wonderful chat with Paul after his talk. He agreed to do an extensive interview for the film and has been incredibly open and generous with his time and friendship ever since. Over the page are Project Drawdown's top 15 solutions, which I will explain in more detail as we proceed.

15 Afforestation

Create forests where there were none before.

14 Tropical staple trees

Eat more foods from perennial trees, such as avocados and nuts (can do!) – as long as no forests are cleared to grow them.

13 Peatlands

Preserve all boggy, soup-like ground, as it holds huge amounts of carbon.

12 Temperate forests

Protect and expand these terrific carbon sinks (and homes to wonderful biodiversity).

11 Regenerative agriculture

Farm in ways that retain carbon or return it to the soil, while giving us healthy food and retaining water. Win, win, win.

10 Rooftop solar

Power your own house or your community.

9 Silvopasture

Integrate trees with pasture or forage crops to raise livestock, and close feedlots!

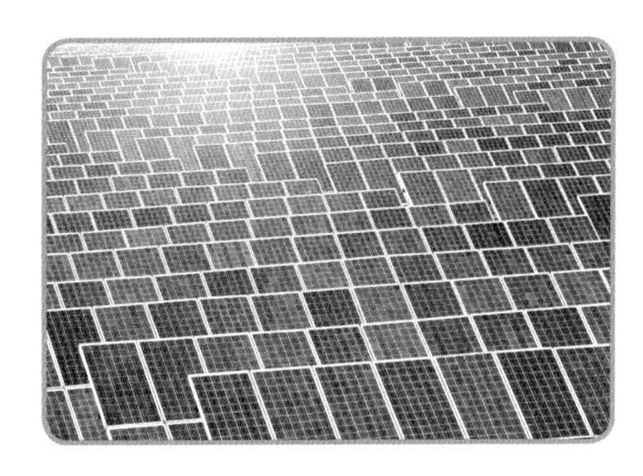

8 Solar farms

Invest in large-scale solar power.

Family planning

Empower women to choose when (or if) they have children.

Education for girls

This leads to empowered girls, who make their own choices around bearing children (more about this on page 168).

Tropical forests

Preserve and restore tropical forests around the world to store carbon and protect the teeming life within them.

Plant-rich diet

Eat less red meat. If you do eat meat, where it comes from is critical.

3 Food waste

Reduce resource waste and the amount of food that reaches landfill by planning meals, storing food correctly, and reusing scraps and leftovers.

2 Wind turbines

Install more wind turbines – the Dutch have known this for years.

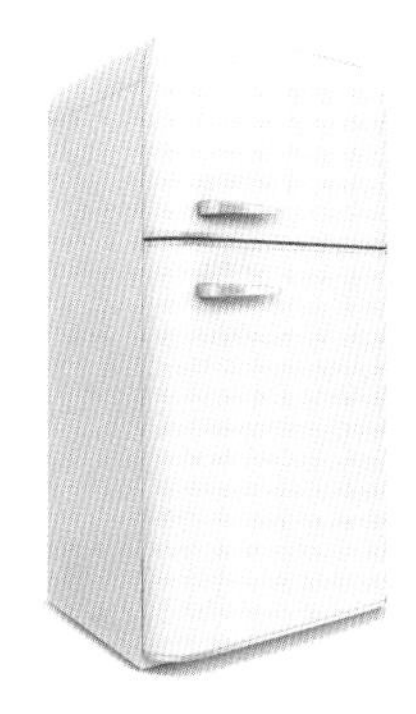

1 Refrigerant management

Phase out all hydrofluorocarbons in air conditioners and fridges as these can warm the atmosphere at a rate 1000–9000 times faster than carbon dioxide. These HFCs are already being phased out through an international agreement.

Although handy to know, some of these solutions can feel out of reach for individuals to tackle. Throughout this book I will end each section with a list of the best things that you can do right now – on your own, as a family, in your office and in the community.

ENE

RGY

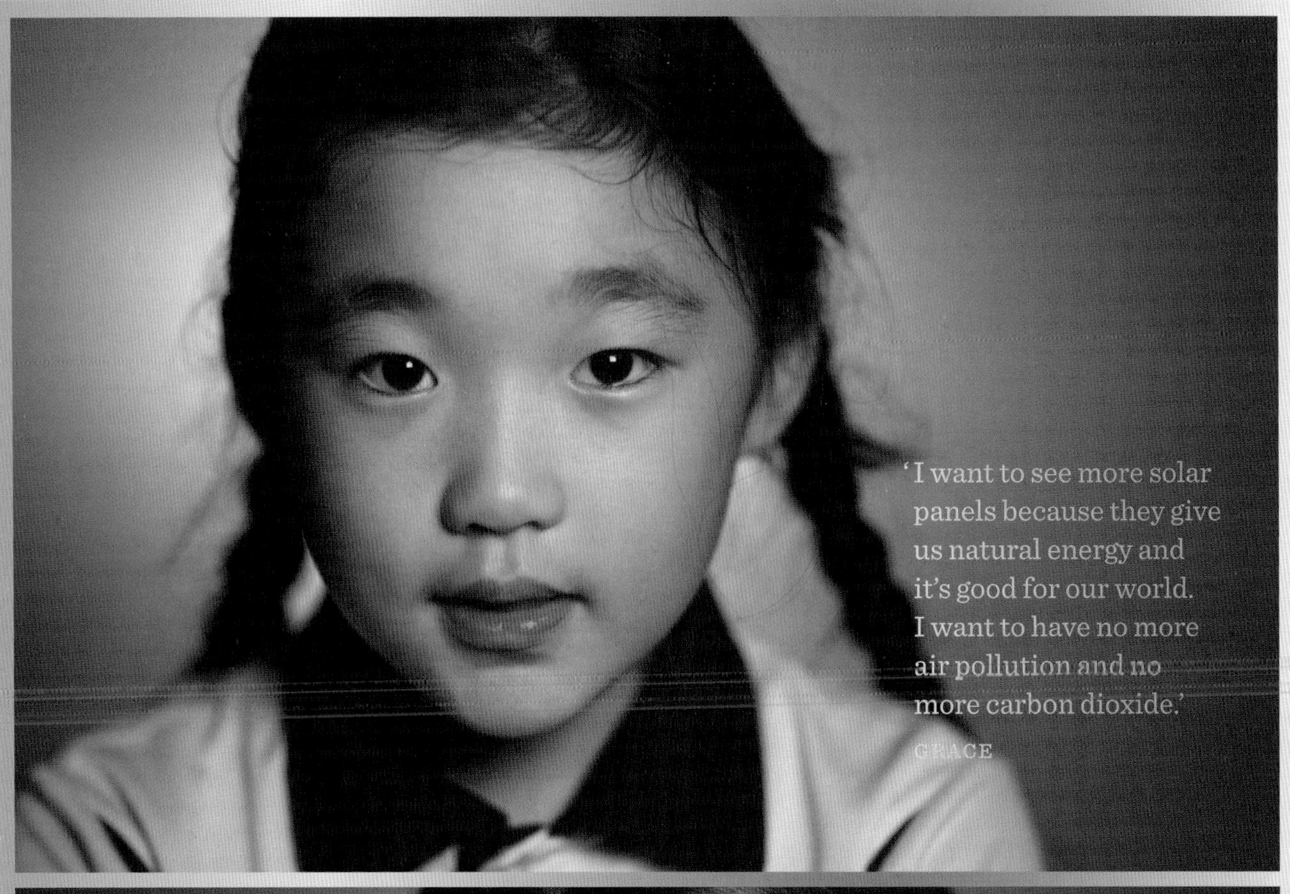
'I want to see more solar panels because they give us natural energy and it's good for our world. I want to have no more air pollution and no more carbon dioxide.'
GRACE

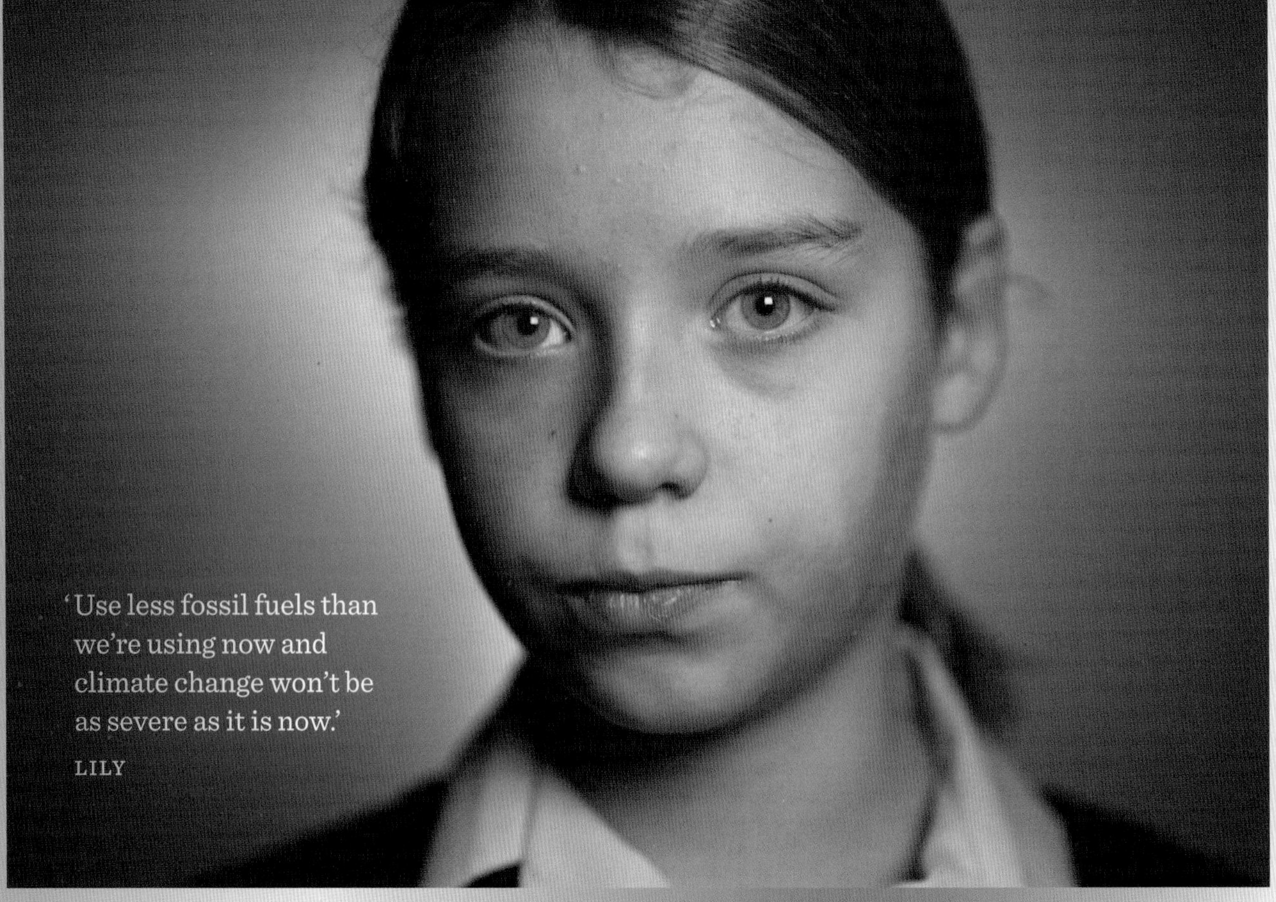
'Use less fossil fuels than we're using now and climate change won't be as severe as it is now.'
LILY

Given that a large chunk of our global emissions comes from making electricity (42 per cent, including heat generation), this seemed the obvious place to start looking for solutions. Most of us know about renewables, though perhaps fewer of us realise we are smack bang in the middle of an energy revolution. According to energy analyst Bloomberg NEF, in 2018 the unsubsidised cost of wind and solar beat coal as the cheapest form of bulk energy generation in every economy except Japan. This includes China and India, where coal has reigned supreme. This is an historic development. Not least because it is happening despite all of the clever tactics being used by those with enormous pools of cash banking on more fossil fuels being extracted (more on that later).

Aside from the focus on renewables, I was interested in finding a different way for us to interact with our energy. Paul Hawken helped focus my research: 'If you combine energy storage with localised generation like wind or solar, then you have an entirely different world than the one we have right now.'

After combing through a variety of innovative renewable energy ideas, I travelled to Bangladesh. It's a shame my luggage didn't join me, but it turns out that two T- shirts, a pair of shorts and a bar of soap is all a travelling 40-something male actually needs.

Bangladesh is a country that understands the need to act fast to protect our environment – a sea level rise of just over half a metre would wipe out 40 per cent of its land. It was sobering to learn that 82 per cent of the costs of global warming are borne by poorer countries, often the same countries who contribute the least to global emissions.

'The wealthiest 7 per cent of the world's people are responsible for 50 per cent of emissions.'

BRIAN TOKAR,
TOWARD CLIMATE JUSTICE

I headed to the remote village of Jote Shouda, a 45-minute life-evaluating flight north from the capital Dhaka, to meet with Neel Tamhane, one of the most switched-on 23-year-olds I'm likely to ever encounter. Neel works for a company called SOLshare, which is at the forefront of the energy revolution. Like my interviews with the kids, meeting someone so young with such passion, talent and understanding of what needs to be done filled me with hope for my daughter's future.

Many households in rural Bangladesh already source energy from a solar panel and battery combo. Their power requirements are much lower than ours in Australia (didn't see one Thermomix), so smaller batteries are adequate. When I first contacted Neel, Bangladesh had more than 5 million solar set-ups, while Australia had around 1.8 million. Thankfully, Australia is now catching up.

Neel and his team at SOLshare provide a special box that allows a home with the solar and battery set-up to connect to another home with the same set-up. Together they form what's called a microgrid – a local network of energy generators. I like to think of it as a way of distributing energy that replicates nature – a series of individual cells that unite to form a larger and stronger organism.

The SOLshare box allows the buying and selling of energy between homes on the microgrid. Even if you don't have solar panels or a battery, the box allows you to buy energy from all the other houses' combined solar panels and batteries, which have now formed a grid.

Neel describes it as a 'water tank of community energy that people can give to or take from. And because people are paying each other for their energy and not giving that money to a large company elsewhere, the money stays in the economy and helps the community. This is bringing people together.'

STEP THREE INSTALL SOL BOX

STEP ONE SOLAR PANELS

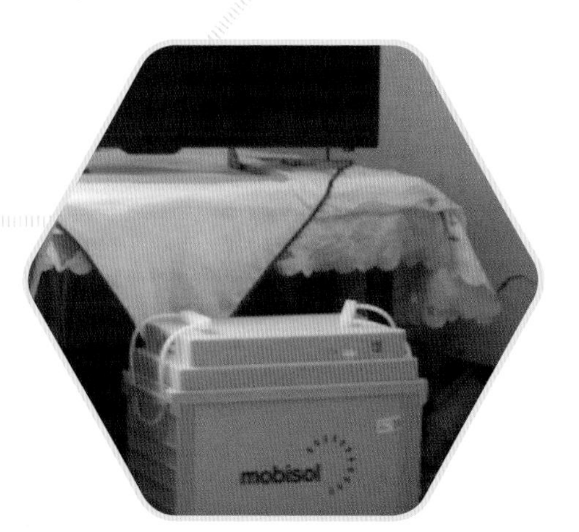

STEP TWO BATTERIES

STEP FOUR THE SOL BOXES CONNECT TO EACH OTHER IN DIFFERENT HOUSES. THIS ALLOWS THE BUYING AND SELLING OF ENERGY BETWEEN THE HOMES.

STEP FIVE WHEN THE HOMES ARE CONNECTED, THE ENERGY IS POOLED TOGETHER TO FORM A MICROGRID. IF A HOME DOESN'T HAVE SOLAR PANELS OR A BATTERY BUT DOES HAVE A SOL BOX, IT CAN STILL BUY ENERGY FROM THE MICROGRID.

After spending a few days with the locals and enjoying some incredible food (and being asked by one lady if I owned any other clothes), the benefits of this technology became obvious. Jote Shouda is a thriving village with a wonderful community spirit. The local businesses are better off, especially the local tailor with his electric sewing machine. The man who runs the shop with the television said they can now watch entire cricket matches instead of seeing just 15 overs before the power cuts out. Families are spending more time together at dinner, plus the children get to finish their homework in the sustained light because even villages connected to the centralised grid lose power between 8pm and 11pm. There's also a new night market, which buzzes with internet cafes, restaurants, fruit and vegetable vendors, and even an electric rickshaw charging station.

Before the solar set-ups, people had to use kerosene lamps for lighting. This is not only expensive but also very dangerous as the toxic fumes are a large killer of women and children in developing nations. It is also estimated that each year 240 million tonnes of carbon dioxide enters the atmosphere from kerosene lamps.

The other major problem is that Bangladesh is very vulnerable to floods and monsoons (Neel told me there were five natural disasters in 2017), which means the centralised power grid often breaks down and people are without power for long periods. Fortunately, villages like Jote Shouda can now keep their power running and, as Neel says, 'aren't reliant on the government for their energy. They are an empowered community.' As the climate disruption continues in the years ahead, this energy resilience will be crucial for all countries.

MICROGRIDS COULD POWER OUR CITIES BY 2040

I interviewed the author and Stanford lecturer Tony Seba, who wrote the book *Clean Disruption of Energy and Transportation*, about this. Tony told me that solar battery prices in the US are coming down faster than solar panel prices (which is saying something) and that by 2021, in many parts of the world, a solar-plus-battery package will be cheaper than the cost of centralised power transmission. 'People will buy their energy systems at the store, at IKEA, Walmart,' he said. 'Eventually, most homes will have a battery the size of a shoebox and energy will be so cheap people won't even notice.' I liked listening to Tony Seba.

Tony told me that by 2040, we will all buy and sell our energy from each other and, due to the low cost and prevalence of solar panels on many surfaces (including our windows and mobile-phone glass), we will be able to donate our excess energy to those who need it. Our energy will no longer be meted out from a centralised utility (dictatorship style) but will be shared peer-to-peer, which is inherently more democratic.

'If you look at newspapers and book publishing 20 years ago,' Tony said, 'essentially someone in a big building decided what information would be consumed or published. They would ship information one way and money would be shipped the other way. It was a centralised architecture. But now every one of us is an information producer, whether on Facebook or Instagram etc. Basically, we are all publishers now.

'The same thing that happened with the web is going to happen with energy. Every consumer, every business, will become an energy publisher. Centralised generation is artificial, decentralised energy is how nature operates – so there's no reason why it shouldn't happen.'

We will, of course, still require larger-scale energy supply, especially for the likes of expanding data centres, but this energy is likely to come from a combination of wind, geothermal, tidal, nuclear (in some regions) and biodigester (which breaks down food waste and other organic matter to create biogas), as well as other sources we haven't thought of yet. Each country is different, each community is different, and they will adapt to the best renewable source available to them.

'Energy is at the core of our civilisation. Soon it will be dramatically cheaper than it is now and this will have an enormous impact on all people and all industries.'
TONY SEBA

Coal workers from Latrobe Valley (clockwise from top): Luke Van Der Meulen, Unit Controller and Union President, Jenny Jackeulen; Ron Ipsen, Staff Operations; Danny Boothman, Assistant Unit Controller.

THE *TRANSITION*

I was lucky enough to spend some time in Victoria's Latrobe Valley, sitting around kitchen tables, dunking biscuits in strong tea, while listening to the stories of the people who were left behind when a local coal power station closed down. These hardworking people provided great lessons for how we can handle many of the changes that lie ahead if we want to heal our planet. Most care deeply about climate change but get infuriated when environmentalists don't stop to consider the impact on their lives. Most had worked in the coal industry for years, often following in their parents' footsteps. They were proud that their jobs literally powered the economy, creating great wealth and opportunities for others. No one had a clue about the impact it was having on the climate.

Governments around the world currently subsidise fossil fuels to the tune of $5.3 trillion a year. That's *$10 million a minute.* (An amount larger than the total health spending of all the world's governments.) As renewables take hold and the need to subsidise fossil fuels is reduced, there is a wonderful opportunity to redirect some of that money to assist the industries most impacted by the transition. The Latrobe Valley people also spoke of the need for community-level support. Yes, financial help and re-employment is important, but the whole community needs to be assisted through the transition. This will happen to more and more coal communities and will extend to other fossil-fuel reliant sectors such as the automotive industry.

Research from the Centre for the Understanding of Sustainable Prosperity (CUSP) in the UK shows that such transitions are possible 'with meaningful, honest dialogue: with workers, communities, entrepreneurs and low carbon businesses of the future; with local and regional governments; and with universities who can ensure support for the skills and jobs of the future not the fossil-fuel past'.

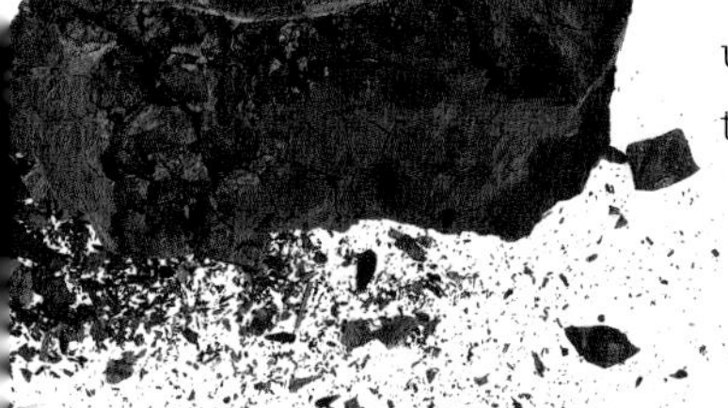

A recent US study analysed the impacts of retraining current coal workers for employment in the solar industry and found that in most cases it would result in better pay for nearly all of the workers. There is a wide variety of employment opportunities in solar, and the industry overall already employs five times more people than coal mining. The only downside is that managers and executives (comprising 3.2 per cent of coal workers) would make less.

- Profitability for US coal power plants has plummeted. Major coal companies are filing for bankruptcy, including the world's largest private-sector coal company, Peabody Energy.
- France has agreed to shut all of its coal power plants by 2023. The UK has agreed to do the same by 2025 (except those using carbon-capturing technology), and Canada by 2030. China recently shut a major Beijing plant and stopped work on 100 new coal power plants.
- The 2020 Tokyo Olympic Games and Paralympics will be powered by 100 per cent renewable energy.
- According to Paul Hawken, 'the price of solar and wind is dropping so fast that the International Energy Agency (IEA) has made inaccurate predictions for the past 19 consecutive years, in each case underestimating the speed of growth.' (Today's amount of renewable energy is 40 times larger than the IEA predicted it would be in 2003.)
- Ireland recently became the first country to completely divest from fossil fuels, passing a law to sell $350 million worth of investments in 150 companies involved in the production of coal, oil, gas and peat.
- Community energy projects (including many microgrids) are springing up all over the world. From India to Scotland to Brooklyn to Daylesford in Victoria, people aren't waiting for their state or federal governments but are taking energy democracy into their own hands.

WHAT YOU CAN DO TO HELP

Please note that most of the actions offered up in these sections of the book relate to things you can do in your home or with your family. While these are important, we also need more people to create change at a higher level. For those wanting to get even more involved with 'The Regeneration', please head to whatsyour2040.com and learn how you can 'Activate Your Plan'.

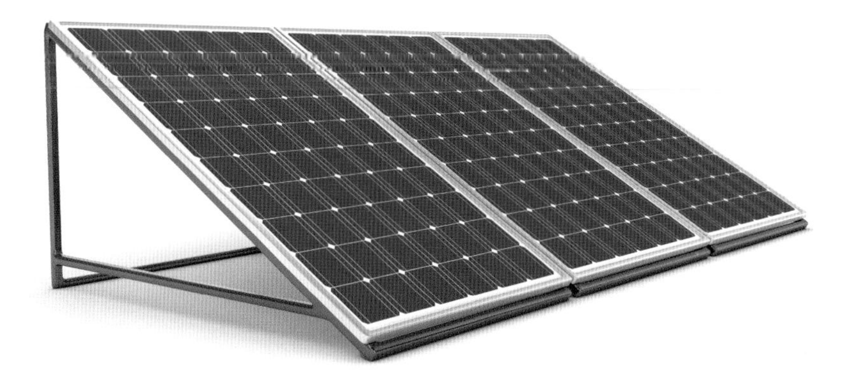

Install solar panels at home if you can

In 2012, the average cost of installing a 5kW system (no battery) in Australia was $10,000. By 2018, it had dropped to under $5000.

If you're a solar newbie, check out the **Renew** website (renew.org.au). It's been promoting renewable energy, energy efficiency and water conservation in Australia since 1980, and its website has everything you need to know about installing a grid-interactive or off-grid solar system.

The **Clean Energy Council** website also has a list of accredited installers: go to solaraccreditation.com.au/consumers.

Once you have solar panels, check if there is a community-owned renewable energy company you can join. Here are a couple of options:

DC Power Co (dcpowerco.com.au)
These guys also install solar panels.

Farming the Sun (farmingthesun.net)
This initiative operates community solar gardens in several areas of New South Wales, and provides research and tools for community groups wishing to develop solar projects.

Choose a 100 per cent renewable energy company

If you can't get solar yourself, consider switching to an energy supplier that sources energy from renewables only, such as **Powershop** (powershop.com.au).

If you're renting and your landlord won't install solar, switch to **Enova Energy** (enovaenergy.com.au). Membership with Enova is portable, so renters can still get their rebate if they move house, as long as they stay with Enova as a supplier. Please note Enova are currently only operating in regional NSW but will be in Sydney very soon. They then plan to operate across the country.

Divest from fossil-fuel supporting companies and banks

Go to **marketforces.org.au** to compare banks, insurers and super funds. We can recommend Future Super and Australian Ethical Super.

Investigate renewable energy for your school

Solar Schools is an Australian/New Zealand-based organisation made up of over 1400 schools. Head to their website to see how you can get involved: solarschools.net.

Investigate renewable energy for your work or community

The **Community Power Agency** (cpagency.org.au) helps people set up community-owned renewable energy. Projects so far include Pingala, Mount Alexander Community Renewables, New England Wind, CANWin and a consortium of groups in the Blue Mountains.

ClearSky Solar (clearskysolar.com.au) provides administrative and web-based support for community renewable energy projects.

Power Ledger provides a peer-to-peer trading platform (similar to Bangladesh) that allows users to trade excess solar power.

Find out what your political party is doing for our environment

- greens.org.au/policies/climate-change-and-energy
- laborsclimatechangeactionplan.org.au
- liberal.org.au/ourplan/environment
- nationals.org.au/our-policies

Improve your energy efficiency at home

Even if you have solar panels, or source energy from a renewables-only power company, it's important to minimise energy use. If you are still using fossil-fuel generated power, then reducing your energy footprint is even more important. Here are some tips that can help:

SPACE COOLING

- Depending on their design, orientation and coverings, windows are responsible for up to 87 per cent of a room's heat gain. In summer, close your curtains or blinds during the hottest part of the day. At night, open curtains and windows to let warm air out and the cool breeze in.
- Choose the most energy-efficient cooling system for your home and climate zone; go to **energy.gov** for some excellent information.
- In summer, set your cooling thermostats to 25–27°C. Every degree below that increases your cooling energy use by 5–10 per cent. Close internal doors and only cool the rooms you are using.

SPACE HEATING

- Insulate your exterior walls and ceiling to prevent 40–60 per cent of heat loss.
- In winter, open blinds/curtains to let the sun in, and close them before it gets dark to keep the heat in – especially while your heater is on.
- In winter, set your heating thermostats to 18–20°C. Every degree above that increases your heating energy use by 5–10 per cent.
- Close internal doors and only heat the rooms you are using.

APPLIANCES

- Look for the Energy Rating label when buying new appliances – taken together, appliances could be responsible for as much as 33 per cent of your energy bill! Consider whether you really need the appliance in the first place (e.g. dishwasher, dryer, extra fridge or freezer etc).
- Run the dishwasher only when full. Scrape plates first then rinse in cold water *only* if they need it (rinsing in hot water negates any energy savings).

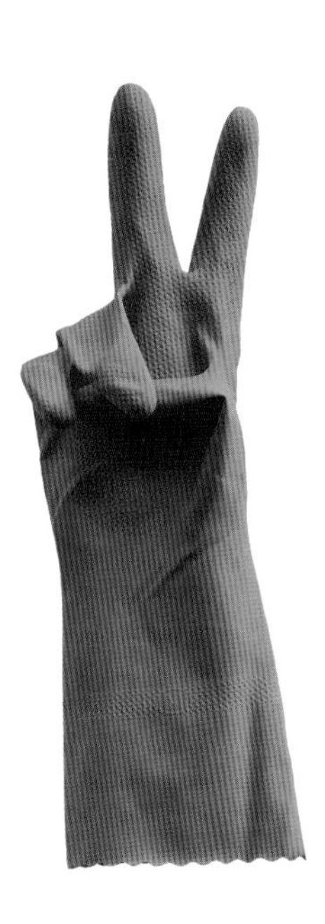

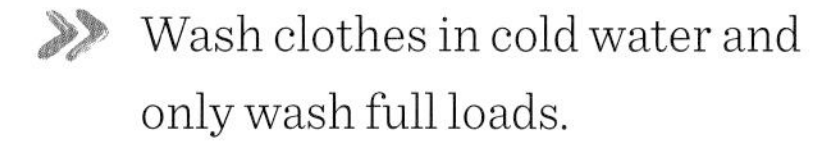

- Wash clothes in cold water and only wash full loads.
- Avoid using a dryer – they use lots of energy. Instead, hang your clothes on the line or near heaters to dry.

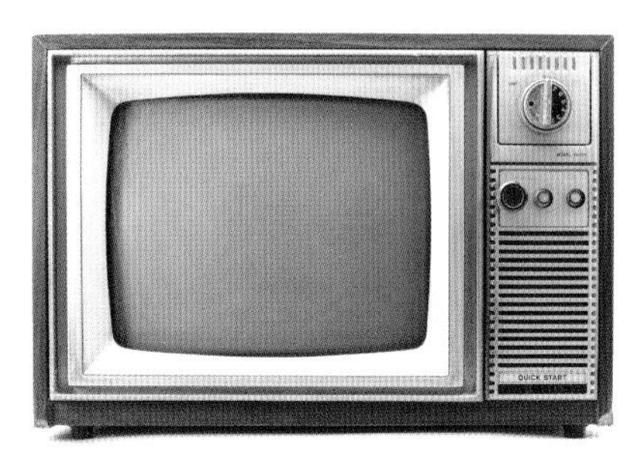

- Turn off all audiovisual appliances (TV, gaming console, computer, printer, phone charger etc) at the wall when not in use, as standby power can account for more than 10 per cent of your household electricity use.

LIGHTING

- Change to energy-efficient light bulbs. Turn off lights when you leave the room.

HOT WATER

- Fit a low-flow shower head and keep your shower time to less than five minutes – this reduces hot water usage by at least 20 per cent.

Despite the positivity of the Bangladesh story, I do have to apply the utopian handbrake for a few paragraphs. As I sat in the airport lounge in Bangladesh (luggage still missing), I discovered that microgrids are currently illegal in some countries, including many states of the US. Their laws stipulate that energy cannot be shared between individuals (peer-to-peer), and must come via a centralised utility provider. These shackles are starting to loosen in Australia, however.

'We really need to wake up to the fact that even the climate discussion has been too narrow. If we only focus on renewables, we're going to be going over the cliff anyway. Happily using our solar power to go off the cliff. Let's really look at this systemically.'

HELENA NORBERG-HODGE,
ANCIENT FUTURES

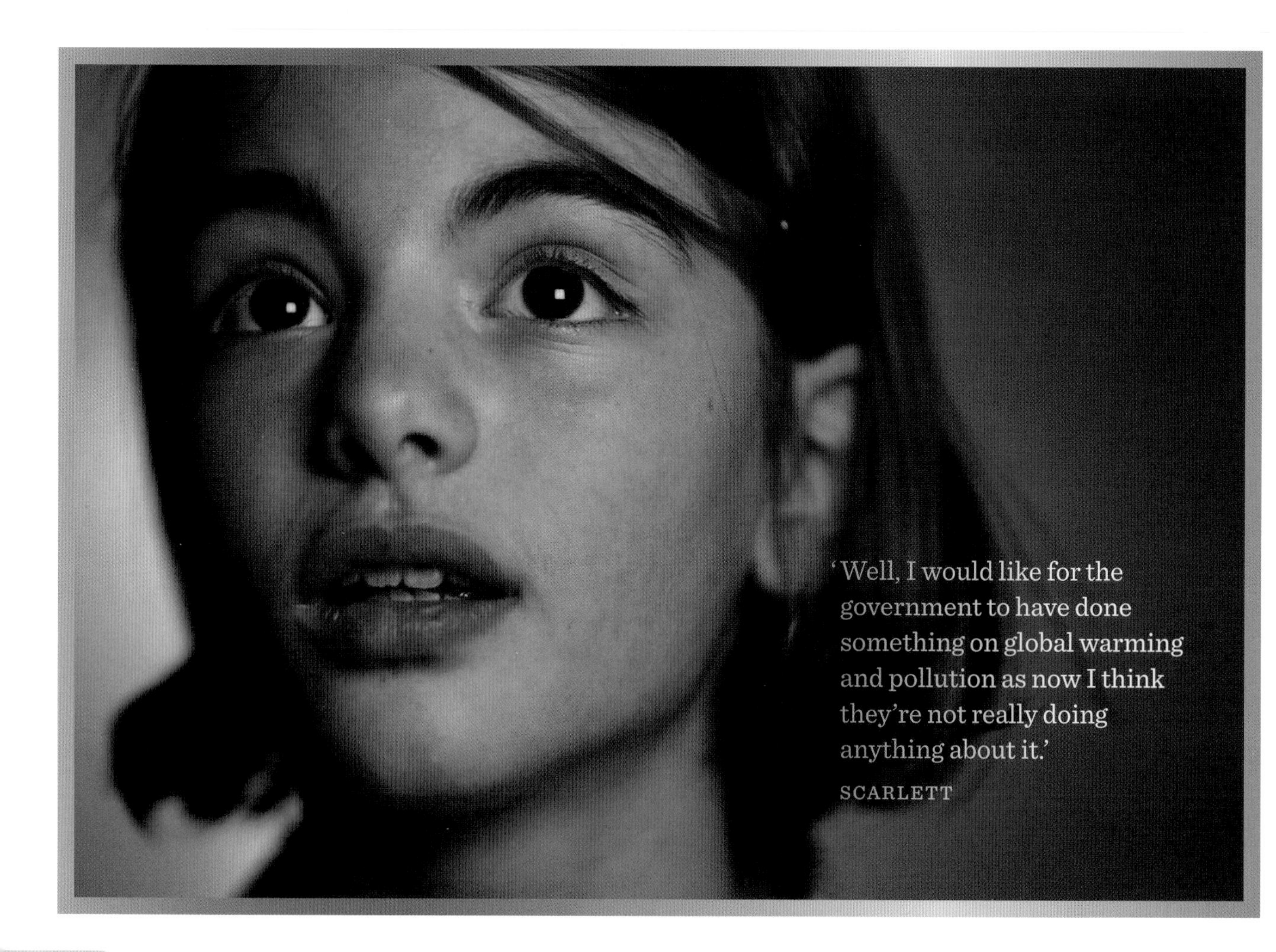

'Well, I would like for the government to have done something on global warming and pollution as now I think they're not really doing anything about it.'

SCARLETT

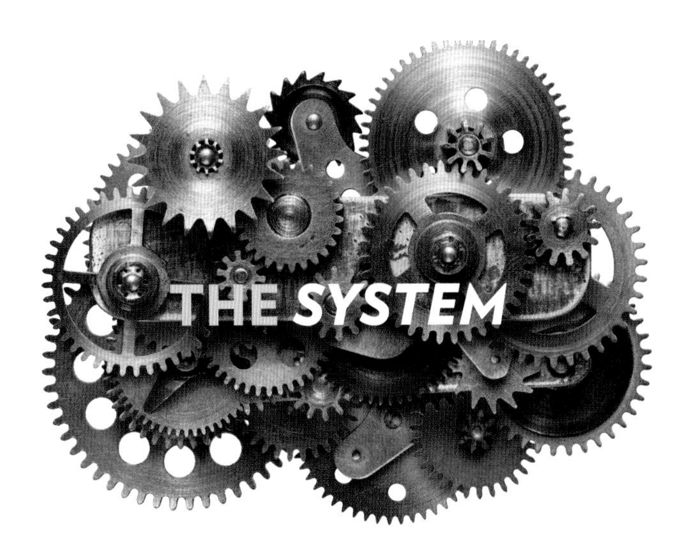

In the current political climate, any discussion of how our economic system operates often descends into a polarised slinging match. You are either a whinging, snowflake leftie or a patriarchy-supporting fascist. Anyone pitching a nuanced tent in the middle can often feel the most alone.

Our current system should be celebrated for its life-saving benefits to billions of people, but it should also be criticised for its neglect of billions of others. What many of its architects could never have predicted, though, was its impact on the planet.

If you had to use a single word to sum up how our system has operated until now, a leading candidate would be 'centralisation'. We've had centralised governments, centralised finance, centralised corporations and centralised energy.

In the late 1970s, after a period of centralised government, it was decided to swing the other way and give more power to corporations and the market. Given the atrocities perpetrated by various governments at the time (at least 170 million people were slaughtered by their own governments in the 20th century), this is understandable. But in order for the market to prevail, environmental protections and workers' rights have been lifted to encourage more economic growth and prosperity. This wealth, we were told, would trickle down to benefit all members of society.

Unfortunately, things haven't quite gone to plan. Not only is our environment suffering enormously (we'll get to this soon), but from 1990 to 2010, 95 per cent of the wealth created remained with just 40 per cent of the population. (CEO salaries in the US have increased by almost 1000 per cent since 1978 while workers' salaries have risen by just 10.9 per cent.) This means that 4. 3 billion people now live on less than $5 a day.

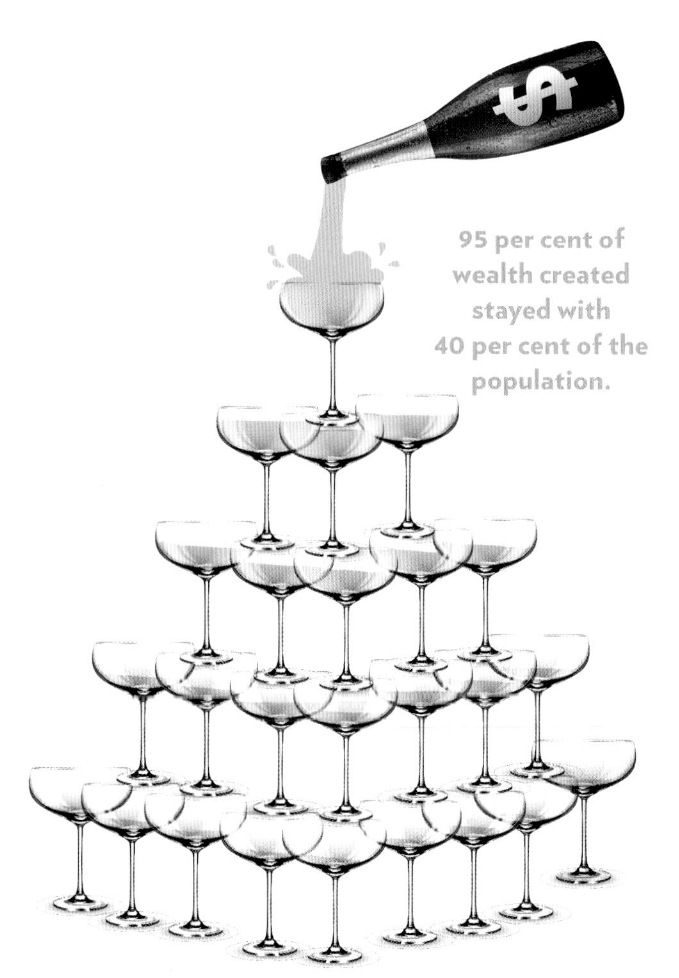

95 per cent of
wealth created
stayed with
40 per cent of the
population.
Trickle Down
Trickle Dow
Trickle Do
Trickle D
Trickle
Trickl
Trick
Just 5 per cent
of wealth made it
to the remaining
60 per cent of the
population.

We have given enormous power to corporations which, as most of us know, now exert a huge influence over our politics. But there is another major area of corporate power that we need to address if we are to make it to a better 2040. There are (often secretive) clauses in trade treaties that allow corporations to sue governments for hindering their ability to make a profit. Now I am all for making a profit, but if the pursuit of that profit overrides the democratic process and threatens the welfare of millions of people and the planet, then it needs to be brought into the light.

One of these trade clauses is called an ISDS (investor-state dispute settlement). A corporation is given the right to sue a government for impacting its ability to make a profit, but the hearing is often held in private, overseen by the World Trade Organization. A revealing example of an ISDS occurred in Hamburg in 2009. A coal power station was to be built on the banks of the Elbe river. After Hamburg's Environmental Authority rejected the project due to its potential to send damaging waste water into the river, the power company sued the government for $1.4 billion in potential 'loss of profit'. The government backed down and the plant was completed in 2014.

'We now have a Swedish nuclear power company – Vattenfall – suing Germany for 4.7 billion Euros because Germany decided, after Fukushima, to phase out nuclear power. We have countries being sued because they want to advertise that cigarettes are not good for you. Countries sued because they want to raise the minimum wage. I think it's really important to understand that most of the people inside these corporations don't even know about this clause.'

HELENA NORBERG-HODGE

There are many examples of the ISDS clause overriding the democratic process, especially around protecting the living world or workers' rights. We have effectively swung from a model of centralised government to one of very centralised corporate power. I'd like to bring this up, though, without falling into an often-used trope that goes something like: 'All corporations are evil.' While a few corporations hover precariously close to the term, the phrase doesn't ring true when the net is cast wider, nor do the people who work inside large corporations resemble anything close to evil (I have relatives who work in the fossil-fuel industry – no goat's blood at Friday night drinks, as far as I can tell). There are more and more examples of companies that are striving to do better. The rise of 'B Corporations' is particularly promising as an emerging corporate and legal structure that binds the company to serve society, not just its shareholders.

Corporations will still be a part of 2040 and so will inequality. The questions we need to ask ourselves are how much power do we want corporations to have, and how big a gap in income inequality do we think is tolerable? Many on my adventure told me that we have already reached the threshold. Research in recent decades has shown that inequality is incredibly significant. Whether a country is rich or poor, a high level of inequality results in lower levels of trust, lower life expectancy, more teenage pregnancy, more drug use, and more people in prison. Democracy also suffers because people don't feel a sense of connection with their neighbours, they don't feel like they're part of a community, so are unlikely to band together to demand things like environmental protections. As one economist you are about to meet told me: 'Getting a more equal society is actually fundamental to so many of the values we are trying to achieve.'

We need to have serious discussions about how we untangle ourselves from this situation. But my experience in Bangladesh shows that the answer may not be a choice simply between 'Big Government' and 'Corporate Rule'. What I witnessed is a new decentralised approach that could, in theory, be welcomed by those on the right for its ability to empower individual freedom and by those on the left for its ability to lift inequality through the redistribution of wealth.

'If we recognise that our wellbeing fundamentally depends on the stability and thriving of this planet, then we will put that at the heart of the economic system we create.'

KATE RAWORTH

I flew to London to meet with the economist and rock-solid human Kate Raworth. When I asked Kate about the top-down, centralised nature of our current system, she shared a story with me about the boardgame Monopoly:

'In 1900, a woman called Elizabeth Magie wanted to create a game that would make people understand just what would happen in life if a few people owned everything. She gave it two sets of rules. The first set of rules, called the prosperity rules, stipulated that when a player buys a property, all the other players get a little bit of income too. The game is won by everybody when the player who started out with the least money has doubled their income.

'The second set of rules, called the monopoly rules, stipulated that when you buy a property, you charge rent to anyone who lands on it and you can use that rent to buy more properties. So the rich get richer, and the poor fall out the bottom. When Parker Brothers discovered this game in the 1930s, they bought the patent from Elizabeth Magie, but they released it to the world with just the monopoly set of rules. So this is the game that kids worldwide are playing today.'

Kate is proposing a new economic framework, designed specifically for the environmental and income inequality predicament we find ourselves in. She calls it 'The Doughnut', and I hope our kids are one day playing a boardgame called Doughnutopoly.

THE DOUGHNUT

AIMING TO MEET THE NEEDS OF ALL *WITHIN THE MEANS OF THE PLANET*

HOUSING

FOOD

We are losing too much biodiversity.

EDUCATION

INCOME EQUALITY

We are clearing too much land and using too many chemicals on it.

Air pollution is costing lives.
ENERGY
HEALTH
WATER
Too many people are stuck in the hole, falling short on life's essentials. We need to get them out.
We are causing climate change and ocean acidification.

Kate's aim is to have society living safely within the soft, doughy ring of the doughnut. Yet we have billions of people stuck in the hole, who are falling short on life's essentials (like food, health and education), while the doughnut itself is threatened from the outside by ecological traumas such as climate change, biodiversity loss, land conversion and ocean acidification. She believes that the solutions we adopt should pull more people inside the doughnut, while ensuring the survival of the doughnut as a whole.

It's a brilliant concept that provides a simple framework to help us reach a better 2040. Almost every government and business believes that constant economic growth is the only way forward. Kate is questioning where that constant growth will take us. Can we keep growing and growing, higher and higher, constantly expanding on a planet with finite resources? There's only so much in the bank and we can't keep overdrawing without consequences.

To put it in perspective, our global GDP (the total value of goods and services in all the world's economies) is around US$80 trillion. This means that, at 3 per cent annual growth, we add the equivalent goods and services the UK produces in a year (around $2.5 trillion worth). That's a year's worth of plastic bottles, forest clearing, coffee cups, e-waste etc. And because the following year's 3 per cent growth is then calculated on a larger figure, the economy gets bigger exponentially. This means that the global economy will double roughly every 20 years. Double, then double again, and again. The Global Footprint Network estimates that we already use around 1.7 Earths' worth of resources every year. Meaning that we are already overdrawing on our bank account.

Kate asks us to consider the point of constant growth if it will destroy the planet and society. She is asking us to shift our focus from simply the *size* of our economy to the *shape* of our economy.

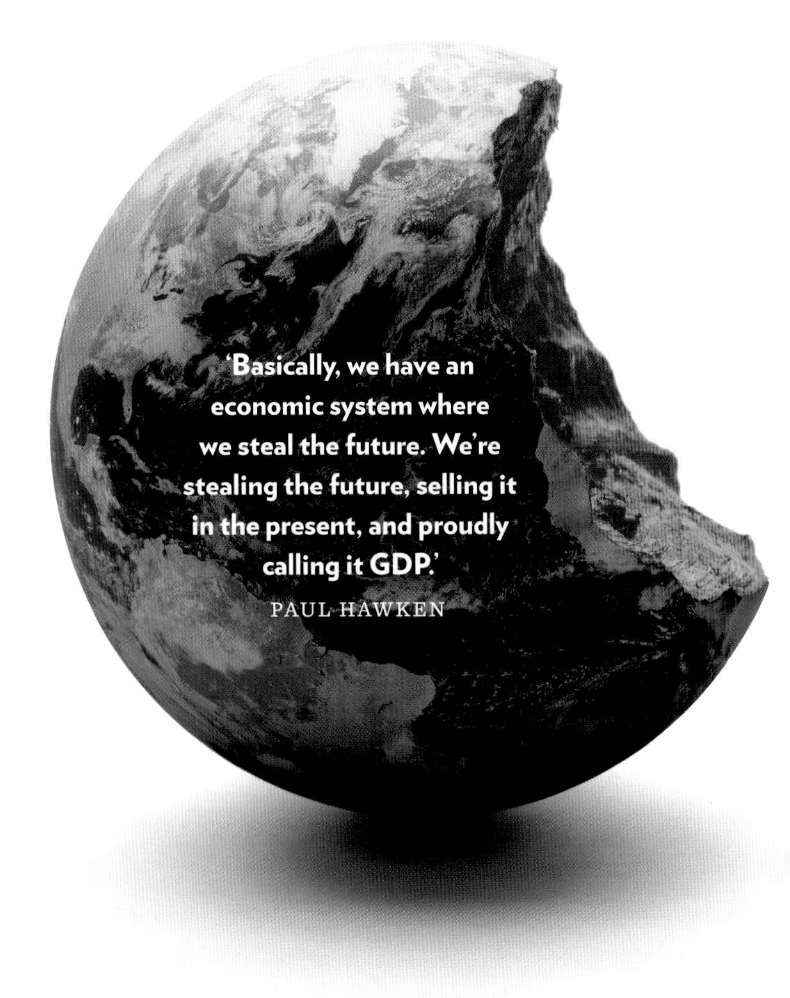

'Basically, we have an economic system where we steal the future. We're stealing the future, selling it in the present, and proudly calling it GDP.'

PAUL HAWKEN

EARTH OVERSHOOT DAY

We are using more resources than the Earth can replenish in a year. We use them all up around 1st August. From then on, we are stealing them from the next year. This diagram shows the resource consumption of individual countries in 2018. Vietnam is the only country close to sustainability; Australia is a long way off.

DEC 21 VIETNAM
DEC 13 JAMAICA
NOV 19 CUBA
NOV 17 COLOMBIA
NOV 6 EGYPT
OCT 28 ECUADOR
OCT 14 ALBANIA
SEPT 25 PERU
AUG 29 MEXICO
JULY 19 BRAZIL
JUNE 10 CHINA
MAY 24 ITALY
MAY 10 JAPAN
MAY 2 GERMANY
APR 21 RUSSIAN FEDERATION
MAR 31 AUSTRALIA
MAR 22 KUWAIT
MAR 18 CANADA
MAR 15 UNITED STATES
MAR 4 UNITED ARAB EMIRATES
FEB 19 LUXEMBOURG

JAN FEB MAR APR MAY JUNE JULY AUG SEP OCT NOV DEC

A team of scientists ran a sophisticated computer model in 2012 that predicted what would happen to global resources if economic growth continued on its current trajectory (increasing at about 2 to 3 per cent per year). It found that human consumption of natural resources (including fish, livestock, forests, metals, minerals, and fossil fuels) would rise from 70 billion metric tonnes per year in 2012 to 180 billion metric tonnes per year by 2050. (A sustainable level of resource use is around 50 billion metric tonnes per year – a boundary we breached back in 2000.)

The beauty of Kate's fried dough and sugar proposal is that it provides a clear set of boundaries within which to operate. She believes there will still be room for growth and abundance but at the point it 'crosses the doughnut's boundaries' a small warning light should flash. I think the solar energy boom provides a great reference for this. China is currently investing $350 billion in solar energy projects. This will provide terrific growth and jobs in the short term but, once installed, the cost of generating solar will fall to almost zero. Kate believes we need to reduce our worship of growth and become more 'agnostic' towards it. We need to allow it to go on a bumpier ride up and down rather than keep it skyrocketing upward, crashing through ecological boundaries.

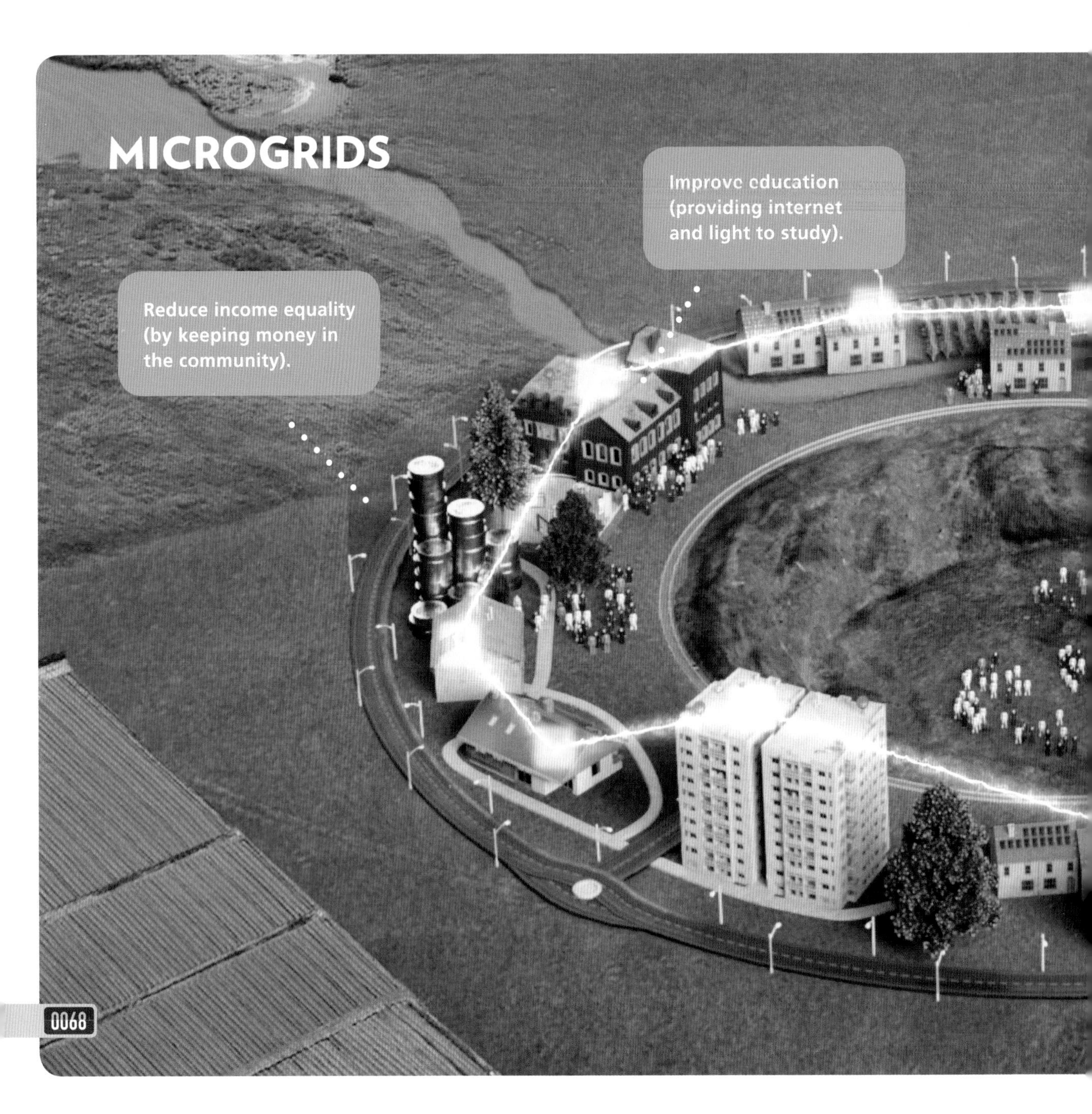
MICROGRIDS
Improve education (providing internet and light to study).
Reduce income equality (by keeping money in the community).

The solar microgrids in Bangladesh are a wonderful example of a doughnut-based solution. The clean renewable energy helps protect the outer boundaries against climate change and air pollution. At the same time the microgrids pull more people out of the hole in the middle by keeping profits within the community (reducing income inequality), lowering kerosene use (which improves health), and providing a constant light at night (which allows families to connect, women to feel safer, and kids to study and complete their education).

Improve health (less kerosene use).
Strengthen networks and community.
'This is our 21st century challenge. Economists and policy makers from centuries before us couldn't even see it. It belongs to us alone. We've got to come up with new ideas of how we're going to meet the needs of all within the means of the planet.'
KATE RAWORTH

TRAN

SPORT

'I would like to see cars
that don't cause pollution,
electrical cars.'
AMBRE

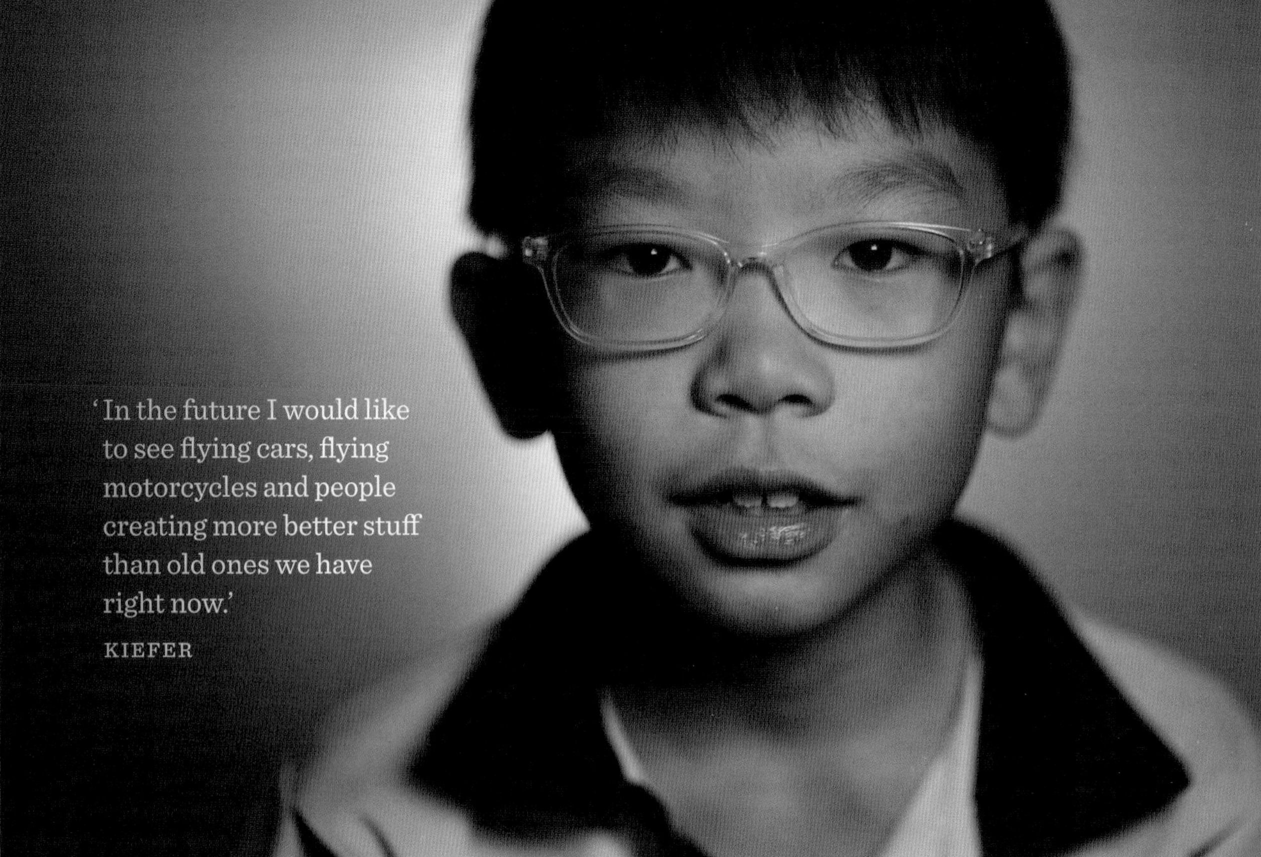
'In the future I would like
to see flying cars, flying
motorcycles and people
creating more better stuff
than old ones we have
right now.'
KIEFER

like my car. It's my podcast-listening capsule, purpose built for really bad karaoke that no one will ever hear. Unfortunately, millions of other people love their cars too, and as a result emissions from road vehicles are one of the largest contributors to our sick planet (20 per cent of emissions in the US alone).

What I don't like about my car is the traffic it often finds itself in. One of the major reasons we moved away from Melbourne was that up to 12 hours of my week were spent in my capsule, and there's only so much of my own singing I can take. By 2040, an extra billion cars are predicted to join the 1.2 billion already on the planet. Imagine what this would do to us mentally. 'Road rage' would morph into 'road haemorrhage'.

My search for alternatives, however, had me a little torn. While I love and embrace certain aspects of technology, I also treat it with caution. I think we can easily fall into a complacent trap of believing that technology will always save the day. As a result, I spent months looking for low-tech ways that we might be able to reduce the number of cars on our roads. Of course there is always taking public transport, walking and riding a bicycle, but, alas, not many tradies I know would get overly excited carrying their tools across town in an 'electric pushbike with sidecar combo', nor would those living in remote areas be keen on subjecting themselves to sporadic bus timetables.

I found myself in Singapore experiencing a very high-tech solution to our voluminous, polluting car problem. It was a ride in an autonomous electric vehicle. That is, a car trip with no driver. It's fair to say I was apprehensive. I have, more than once, asked Siri to 'call Zoe' and ended up with a recipe for calzone, so to now trust one of Siri's cousins with my life in traffic had me a tad nervous. Of course, as I soon discovered, the technology in these cars makes Siri look like a Pac-Man arcade game from a 1980s milk bar. The car uses lidar technology, which reads almost every object, still or otherwise, around the vehicle for a 50-metre radius and can detect a pinkie finger twitch on a human hand 30 metres away. And this technology is improving every year.

But the driverless car solution is only one link in a disruptive chain that the author and Stanford lecturer Tony Seba has been describing to governments and business leaders for the last decade. Spending time with him opened my eyes to the possibilities of this technology – and how it could dramatically transform our culture, our cities and our environment.

According to Seba, 'A disruption occurs when technologies converge that create new markets, products and services while radically transforming existing industries.'

Clear recent examples include the effects of Uber and Lyft on the taxi industry, and of Netflix on DVDs. Another, widely referenced example is the story of Kodak film. The company foresaw the digital photograph revolution but failed to see the convergence of other disruptive factors, namely the rise of mobile phone cameras and the explosion of photo-sharing on social media platforms. Kodak plummeted from a record year in 2000 to bankruptcy court in 2004. Tony Seba believes that a similar disruption may be coming for the car industry and for mobility as we know it.

The approaching convergence is made up of electric vehicles, autonomous (self-driving) vehicles and the ride-share model of Uber, Lyft and Shebah. I will separate the three and then see how they might create the disruption when combined.

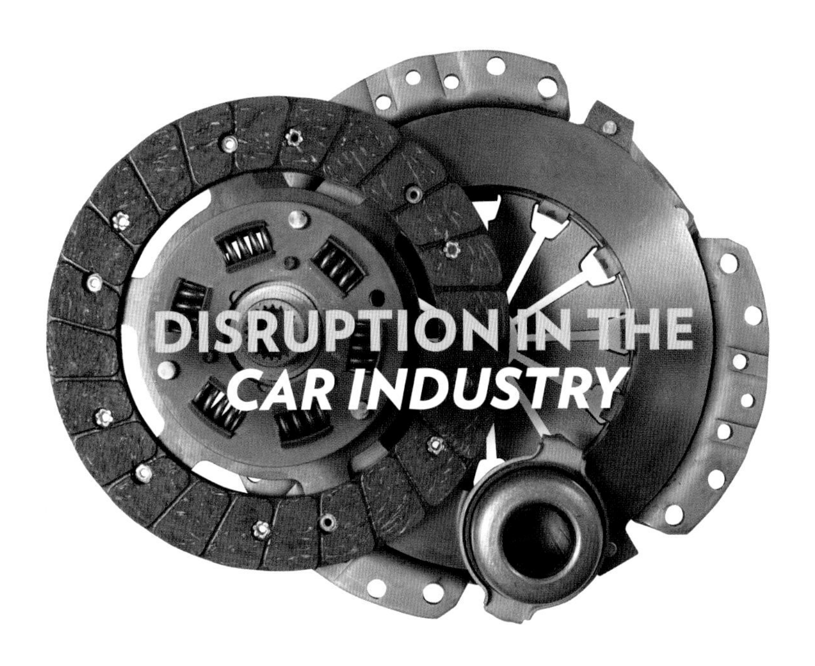

DISRUPTION IN THE *CAR INDUSTRY*

ELECTRIC VEHICLES

Electric cars are getting cheaper and are also a superior product. Not only do cars like the Tesla or Nissan Leaf consist of around 20 moving parts, compared to a combustion engine's 2000 moving parts, they also cost around 90 per cent less per mile to run than a petrol/gasoline car (based on US figures). In Norway, nearly 50 per cent of new cars are electric due to the government removing taxes, plus giving free parking and toll-road use to owners in cities. (Here in Australia we slap a luxury car tax on electric vehicles, making them ridiculously expensive – insert face-palm emoji.)

RIDE-SHARING

In just 10 years of operation Uber is creating more bookings in the US than the country's entire taxi industry. In fact, 20 per cent of all miles travelled in San Francisco are now in either Uber or Lyft vehicles. Other companies are now entering the market, including blockchain groups such as Arcade City, which cut out the intermediary to create even cheaper, peer-to-peer ride sharing.

'In 2015, we lost 1.2 million people on our roads. That is the equivalent of 12 jumbo jets full of passengers crashing every day. Our data shows that driverless EVs will reduce that number by around 98 per cent.'

HUSSEIN DAI
SWINBURNE UNIVERSITY OF TECHNOLOGY

AUTONOMOUS VEHICLES

More and more companies are emerging in this space and the technology is rapidly improving. The company Zoox now moves people around San Francisco in driverless, square-shaped vehicles (although a driver still has to be behind the wheel in case of emergencies). If regulation allows, driverless vehicles or pods could operate around our cities with huge savings.

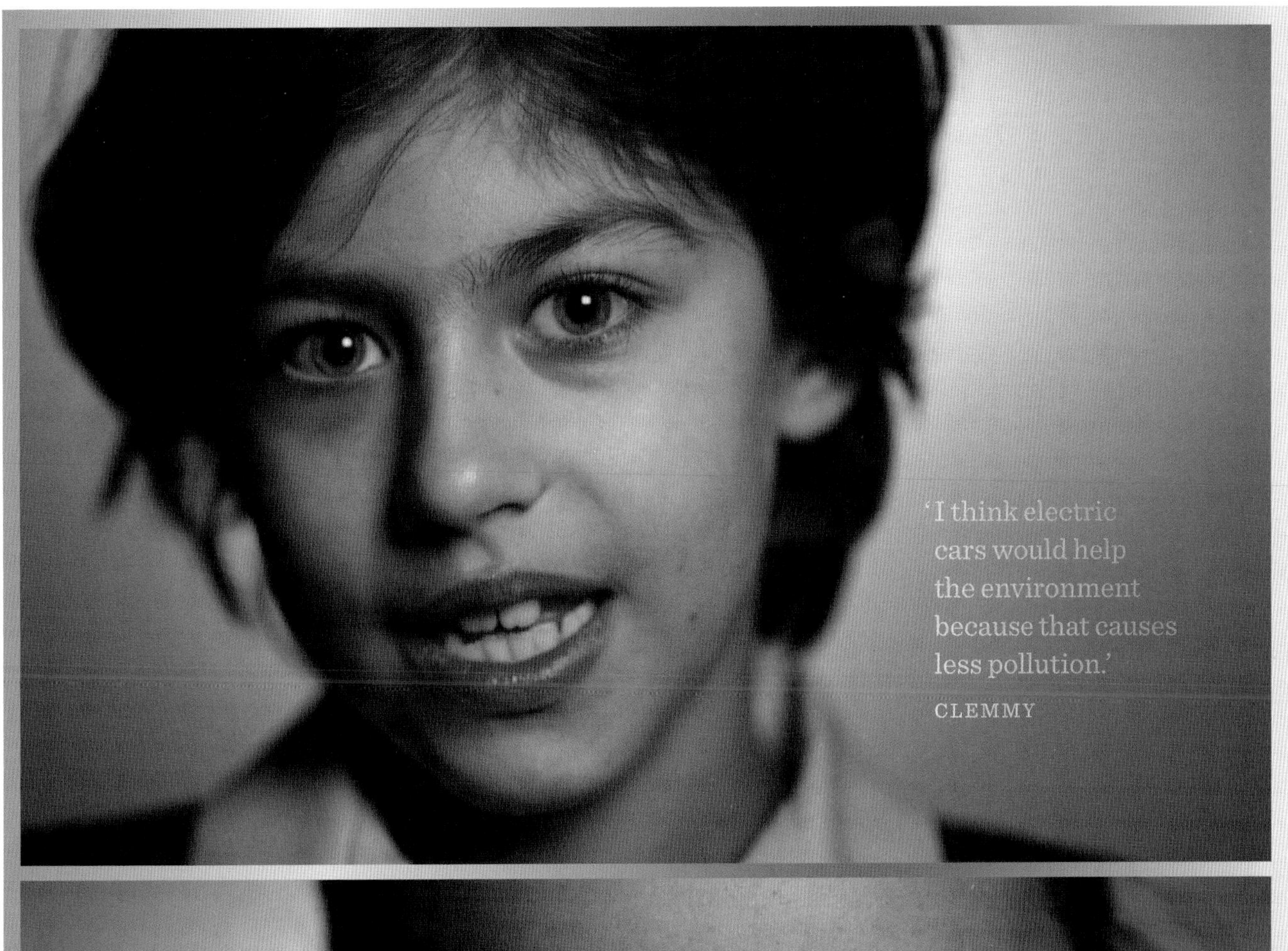

'I think electric cars would help the environment because that causes less pollution.'

CLEMMY

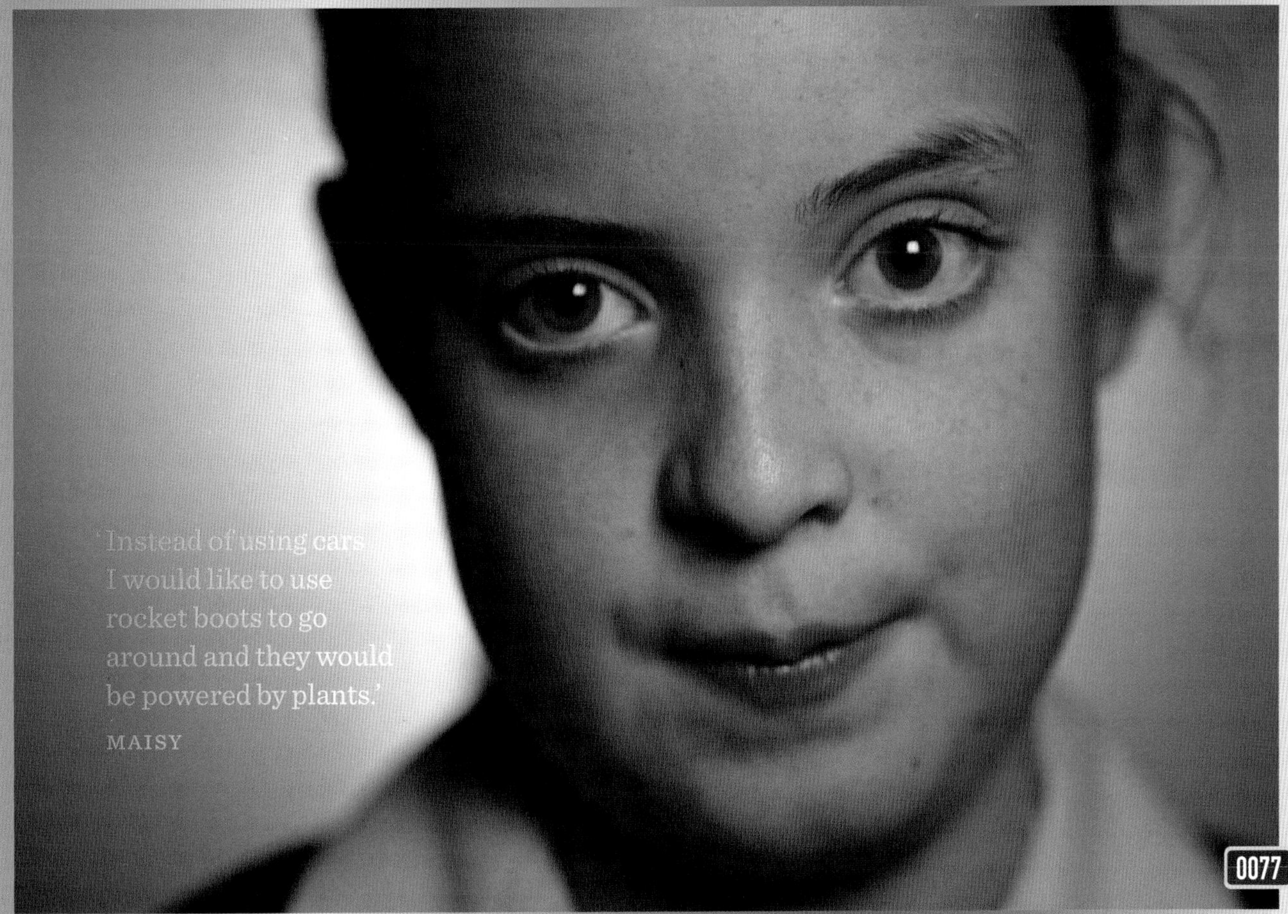

'Instead of using cars I would like to use rocket boots to go around and they would be powered by plants.'

MAISY

According to Tony Seba and his team at RethinkX, if you combine cheap electric vehicles with driverless ride-sharing (which becomes cheaper than current ride-sharing because there is no driver to be paid), then you get a dramatic disruption to the automotive industry. The cost of transportation per mile/kilometre for an individual will go down 10 times relative to buying a new car.

'Somebody is going to need a new car,' said Tony, 'and is going to think, do I want to spend $50,000 over the next five years just owning the car, paying insurance, registration, parking etc, or do I want to not own a car and save $45,000? It's a no-brainer. I mean, where is your car now? It's parked somewhere. Around 96 per cent of the time it is parked or unused. It's a stranded asset that you are paying a lot for.'

Tony is suggesting that, in the future, fewer people will own cars and will instead rely on fleets of cheaper, readily available, driverless cars to run errands and get them from A to B. This is a very compelling argument that many other experts have reinforced, but I must admit that the image of robot cars driving us around does make me feel slightly uncomfortable. As does the idea of a small group of companies owning all of our transportation (Google, Apple and Uber are already investing in driverless transportation fleets). Even after experiencing a ride in a driverless car I wasn't sure if this solution would even make the cut of the film.

I then spent an afternoon with two different academics. One was the anthropologist Genevieve Bell from the Australian National University, and the other was Hussein Dai from Swinburne University of Technology in Melbourne. Genevieve's work looks at behaviour change around new technologies, while Hussein's work focuses on what an autonomous vehicle disruption could do to our cities.

Hussein told me that in the same way that 'on-demand' movie or music services, such as Netflix and Spotify, have led to us owning fewer actual DVDs or CDs, on-demand transport could dramatically reduce the number of cars we own, especially in cities. It could also mean we'd require almost no parking spaces in our cities as the on-demand vehicles would be travelling around in fleets picking people up and dropping them off. There could even be cafe-type vehicles where you could sit with others and drink coffee while getting somewhere. Or 'office pods' where board meetings could take place in transit, comfortable 'library vehicles' and even 'holiday pods' that could take you to a coastal destination and park with a view of the ocean to wake up to. I used to get up very early every morning and drive an hour to work when we lived in Melbourne. The thought of taking a 'bed pod' to work is very appealing (although I suspect such a pod could be used for 'other purposes' – and who would change the sheets?).

With fewer vehicles, a positive outcome could be that our cities have huge amounts of space freed up over the next 20 years (especially cities like Los Angeles, which are predominantly made up of parking lots and roads). Imagine if we decided to replace car lanes and car parks with bike paths and public parks, or more areas for sports or art, or urban food projects or more affordable homes? We would have significant societal decisions to make about what our 'new' cities would look like. At the same time, we would be electrifying our transportation and reducing the number of cars, which translates to drastically reducing our emissions and keeping a lot more fossil fuels in the ground.

These possibilities are the reason this solution stayed in the film. I couldn't find any other way to reduce congestion, reduce the resource use of building an extra billion cars, improve our environment and potentially reinvigorate a feeling of community in our cities. This solution, like the microgrids in Bangladesh, could really fatten Kate Raworth's doughnut.

> **'Two-thirds of Los Angeles is parking and roads. That city is going to have to make a choice. Do they want a giant parking lot or do they want the beauty of a green city?'**
>
> TONY SEBA

Once again, however, it is time to apply the utopian handbrake (would driverless cars even have a handbrake? Perhaps it'd be an emergency lever to override a rogue robot). There are issues to be resolved: the loss of millions of automotive-related jobs around the world; regulation; who is protected in an emergency (the passenger or pedestrian?); and the potential for large corporate purchases of the newly available space in our cities. All very legitimate concerns.

The jobs argument will probably receive a huge amount of attention, and justifiably so. Apart from automotive manufacturers and drivers being put out of work, there is also the impact on mechanics and other 'after market' industries when cars go from 2000 moving parts and regularly needing a service, to 20 moving parts and being handled predominantly by computer technicians. I recently watched a program from the 1960s, in which schoolchildren were interviewed about what they thought the world would be like in the year 2000. They nearly all said that work would be done by robots or machines and that no one would have jobs. It is a popular narrative but underestimates the creativity of humans and their brilliant ability to find ways of adapting and surviving. There's no doubt driverless cars will have an impact, but the fact that Australia (like many nations) still has huge areas without wi-fi means autonomous trucks carrying goods from Adelaide to Darwin will not be happening any time soon.

'There's never been so much good work to be done than right now. My motto about regenerative development is: "Let there be jobs". Because really what we're talking about is creating those kind of jobs. Full employment in the restoration and the regeneration of the Earth, which has so many benefits. They're uncountable, really.'

PAUL HAWKEN

While making the film, I often asked taxi or Uber drivers about this topic. Many told me that they would be happy to relinquish their driving job if they could be retrained in a new career. They were quite excited by the idea of helping to rebuild a better, cleaner city for their children and wanted a job with 'purpose'. (I can relate. I spent five years as a Pizza Hut delivery guy, which involved many humiliating door-knocks at parties where a voice from within would shout to the door-opener: 'Is he cute?' There often followed a brutal three-second pause while the door-opener stared at me, then replied: 'Too many zits.' The smell of ham and pineapple still triggers floods of anxiety. I would have happily taken any job that regenerated the planet over that Hawaiian-scented pain.)

It would be great to start having these discussions now about how we handle this automotive transition. And, importantly, how best to redirect some of the soon-to-be redundant $10 million-per-minute fossil-fuel subsidies.

All the experts I spoke to told me that, whatever transpires, cities will look very different by 2040. They all said that transportation is heading rapidly towards electrification. We could even have short-haul aeroplane flights running on electric batteries by the late 2020s. We could see electric ships (using the batteries as ballast) carrying goods around the world, and an increase in superfast trains (potentially Hyperloops) connecting more and more cities from 2030 onwards.

HOW CLEAN TRANSPORT COULD BRING PEOPLE INTO THE DOUGHNUT AND IMPROVE OUR ENVIRONMENT
Space for urban food.
Increased income from the jobs created redesigning cities.

MORE REASONS FOR
HOPE

- India aims to sell only electric cars by 2030 (Norway and the Netherlands are aiming for 2025).
- Volvo has announced that it will make only electric vehicles from 2019.
- Rooftops in commercial areas of France must now have either rooftop gardens or solar panels.

Improved health from more walkways and bicycle paths in city centres.
Reduced air pollution.
Vote
Slowing down or reversal of climate change.

- Research has found that if global cycling rates can rise from their current level of 6 per cent (it's only 1 per cent in the US) to around 14 per cent, then urban carbon emissions will drop by 11 per cent. Boosting walking rates would have similar benefits.
- In Seattle, Mayor Jenny Durkan recently allocated more than $46 million in the city's budget for cycling and pedestrian improvements, and pledged to grow the public transit service by 30 per cent.
- Residents and workers across three towns in Singapore will use driverless vehicles as part of their daily commute by 2022.

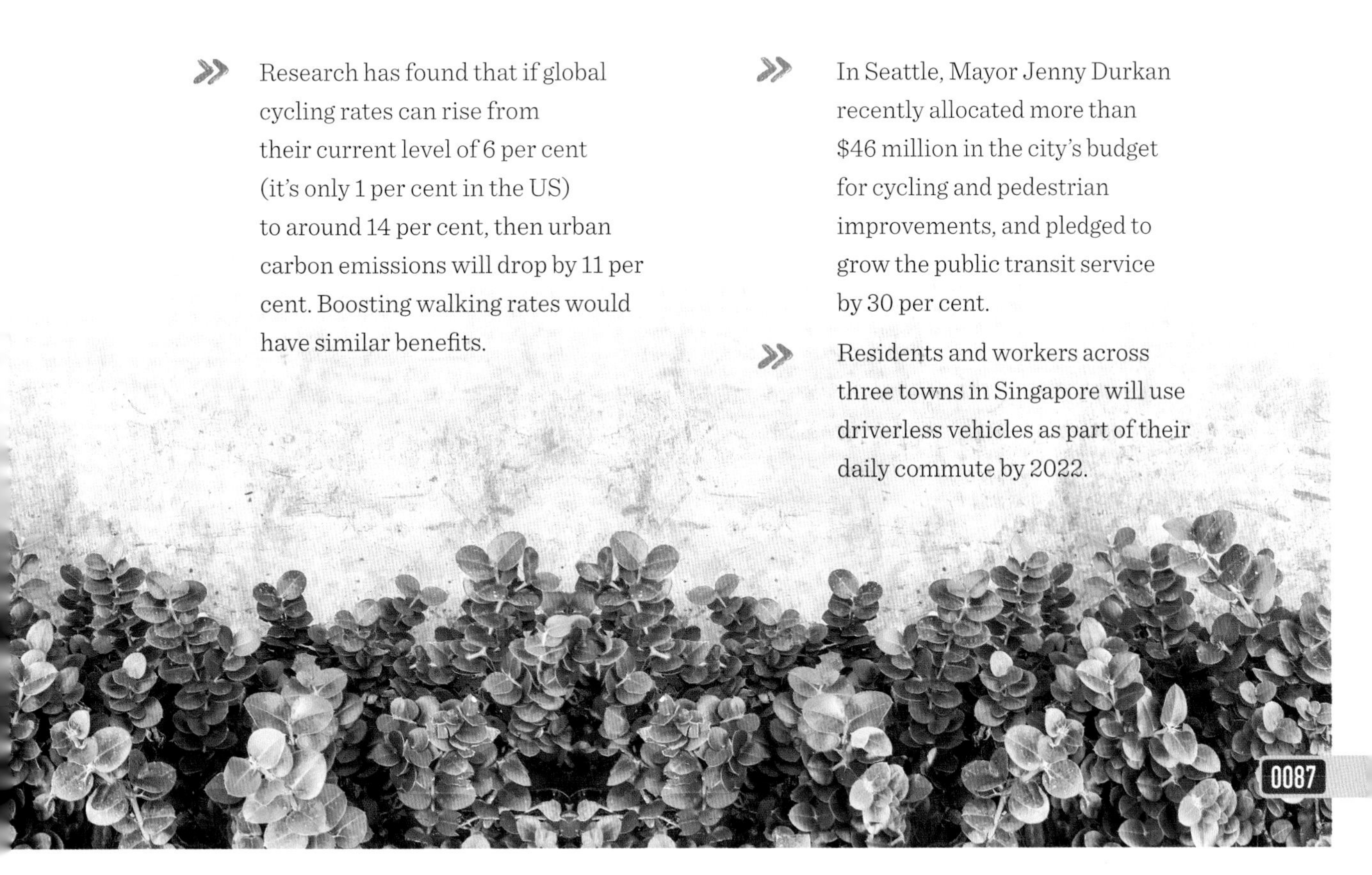

WHAT YOU CAN DO TO HELP

It's important to remember that you can't help but be a hypocrite when it comes to the transport sector right now. There just isn't another way to fly and I certainly can't afford an electric car in this country. But, as we know, the worst thing we can do is feel guilty about our actions around climate because guilt just shuts us down. So unless you are prepared to kayak to visit family or friends in Europe, there is no point beating yourself up – although there are ways to dramatically reduce your carbon footprint while doing so. If enough of us did the following things more often, it would make an enormous difference:

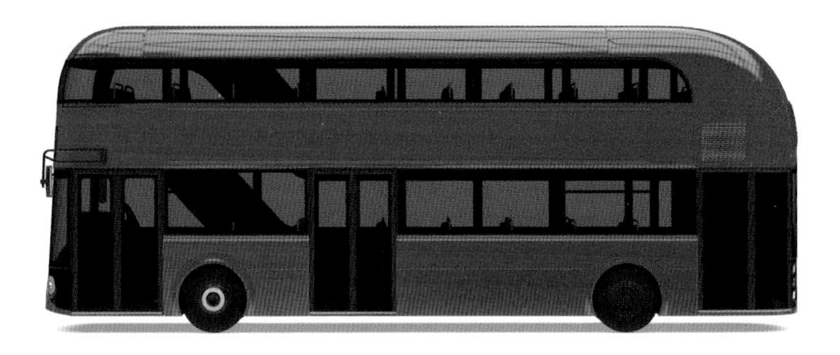

Stay informed

I can highly recommend **The Driven** (thedriven.io) and **Renew Economy** (reneweconomy.com.au) as sources of all things clean energy. Their positive stories of our energy revolution lift the spirits each day.

Use public transport or your own steam

Use public transport (bus, train, tram) or your own steam (bike, scooter, skateboard, inline skates, your own two feet) to get to work at least one day a week, if you can. Statistics show that if everyone in the US stopped driving for just one day a year, it would save 3.5 million tonnes of carbon dioxide. (I recently purchased a second-hand electric bike on Gumtree. Love it. Feels like I'm cheating and not really riding to work, but then I don't rock up dripping in sweat and needing a shower.)

Ride-share

Ride-share with a work colleague at least one day a week. Again, this means one person is not driving for a day. It matters. Next time you're driving somewhere, take a look at how many people are by themselves in their car listening to the radio or a podcast, not even singing karaoke! Such a waste. May as well be ride-sharing.

Car-share

If you live in a city, do the maths on your car ownership. A friend recently sold her car because she calculated that it was cheaper to use Uber, trains and her bike rather than paying for registration, insurance, expensive parking and repairs. By a fair margin too. 'More money for beers' was her reason. She often uses a car-share company or a hire car for errands around the city or weekends away with friends.

Car-share companies include **GreenShareCar**, **PopCar**, **GoGet** and **Flexicar**, and peer-to-peer car-shares include **CarNextDoor** and **DriveMyCar**.

Reduce your air travel

Air travel makes up almost 2 per cent of global emissions (private jets are a major culprit), but until electric or hydrogen-powered planes emerge (or ones powered by algae, a technology that is in development), air travel is still one of the most damaging ways we can impact the planet. An economy return flight from Sydney to London uses the same amount of greenhouse gases as running a standard air conditioner for 7,133 hours. Ouch. Business class uses triple that amount due to more room being taken up by a single passenger. I often try to Skype for a progress meeting and try to fly only when I have to be physically present in a room.

Reduce your travel footprint

If you are going to fly, there are other ways to reduce your carbon footprint:

- Try to find hotels or Airbnbs that use renewable energy. The Global Footprint Network says this can reduce your travel footprint by up to 48 per cent.
- Hire or rent an electric vehicle when you land. Better still, use local public transport or hire a bike to get around. It's a great way to immerse yourself in a new city.
- Eat local food.
- Choose 'carbon offsetting' when you book your flights (it will pop up as on option when booking – roughly $2–3). This helps, yet less than 10 per cent of Australian travellers choose this option.

BLOCKS TO PROGRESS: *VESTED INTERESTS*

In 1989, the then UK Conservative Prime Minister, Margaret Thatcher, addressed the UN and spoke of how the greatest threat to mankind wasn't war, but something else: 'What we are now doing to the world, by degrading the land surfaces, by polluting the waters and by adding greenhouse gases to the air at an unprecedented rate – all this is new in the experience of the earth. It is mankind and his activities which are changing the environment of our planet in damaging and dangerous ways . . . The result is that change in future is likely to be more fundamental and more widespread than anything we have known hitherto.'

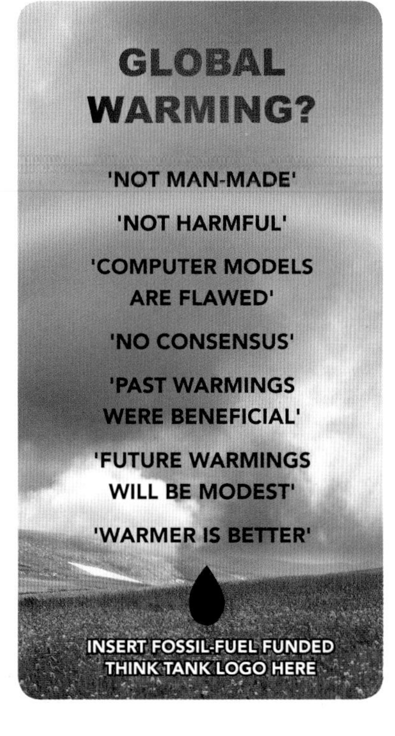

GLOBAL WARMING?
'NOT MAN-MADE'
'NOT HARMFUL'
'COMPUTER MODELS ARE FLAWED'
'NO CONSENSUS'
'PAST WARMINGS WERE BENEFICIAL'
'FUTURE WARMINGS WILL BE MODEST'
'WARMER IS BETTER'
INSERT FOSSIL-FUEL FUNDED THINK TANK LOGO HERE

A few months later, George Bush Senior, the then US Republican President, also spoke to the issue: 'We all know that human activities are changing the atmosphere in unexpected and unprecedented ways. Much remains to be done . . . And together we have a responsibility to ourselves and to the generations to come, to fulfil our stewardship obligations.'

But just as these world leaders were bringing the dangers of burning fossil fuels into the public consciousness, companies like Shell and Exxon were working on a public awareness campaign of their own. As leaked internal videos and documents now show, these companies were well aware of the damage their products were doing to the planet and so decisions were made to begin a denial and misinformation campaign that rivals Big Tobacco's.

Two prominent figures in the denial movement are Charles and David Koch, who have made obscene wealth in the fossil fuel and chemical industries. In 1980 David Koch ran as the far-right US Libertarian Party's vice-presidential candidate, coming from a position even further right than the eventual President, Ronald Reagan. The Libertarian's platform called for the abolition of all government healthcare programs, social security, all income and corporate taxes, the Environmental Protection Agency, the FBI, the CIA, any laws impeding employment (including child labour laws), the Food and Drug Administration and, not surprisingly, an end to the prosecution of tax evaders.

The Koch brothers have always had a deep mistrust of government, which is kind of understandable when you learn that not only was their father an oil manufacturer in Stalin's Russia and Hitler's Germany, but also that the boys were raised by a Nazi-sympathising nanny who ran the household with authoritarian rigour.

When David lost the VP nomination, the brothers decided that politics was not the way to influence American people. They had to go much deeper and change the thinking and culture. They began pouring money into establishing think tanks that supported their views, funding advertising campaigns, front groups, lobbyists, academic programs, and books: anything to glorify individualism and the 'free market'. In 1978 Charles Koch declared, 'Our movement must destroy the prevalent statist paradigm.'

Unfortunately for much of the living planet, they have been very good at telling their story. With so much of their money tied up in fossil-fuel extraction, climate change denial has been a lead character in their story, and they have poured millions of dollars into such campaigns. This is delaying crucial action being taken.

Climate change denial has become a powerful case study for agnotologists, a word that is being used a lot more in recent years. Agnotology is the study of how ignorance is spread through a society. But before I outline some of the tactics that have been used, I want to mention that people's beliefs about climate change are incredibly complex. Some people genuinely believe that climate change isn't real or happening, and, for them, facts or evidence play only a tiny role. I don't think everyone has simply been misled by a few vested interests. A person's stance on this topic can be informed by many factors: their political and religious views, their trust in or mistrust of authority or science, or the influence of websites or news channels that align with their values. So before thrusting the following text into the face of a sceptical brother-in-law or climate-denying work colleague, it might be worth keeping that in mind.

Ten years ago, I was firmly in the sceptical brother-in-law's camp and was dubious about the legitimacy of climate change. I'd been guzzling down the 'it's a hoax' flavoured Kool-Aid. Learning about the tactics used to deny climate change has allowed me to see things in a new way. Ultimately, it doesn't matter if any of us believe in climate change or not; it's already happening.

THE CLIMATE DENIER'S HANDBOOK:

Seven Ways to Influence People's Views

1.

SPEND BILLIONS OF DOLLARS ON PROPAGANDA.

Robert Brulle from Drexel University painstakingly combed through the tax returns of think tanks and lobby groups and calculated that vested interests spend a total of almost $1 billion a year on climate denial propaganda. The Union of Concerned Scientists found that between 2005 and 2008 alone, Exxon spent $8.9 million and the Koch brothers spent $24 million on the dissemination of climate misinformation. The so-called 'whinging lefty greenie groups' cannot bring balance to the media narrative when competing with that level of funding.

2.

SET UP A BUNCH OF ORGANISATIONS TO APPEAR LARGER IN NUMBER.

It is now well known, thanks to the terrific work of Naomi Oreskes and others, that Exxon set up and funded a range of companies and organisations under different names to give the impression that there were lots of climate denial groups. A power company in New Orleans was recently caught paying actors to attend a council meeting (with placards) pretending to support a new natural gas plant. Sneaky buggers.

3.

MEET REGULARLY TO DISCUSS TACTICS.

The Heartland Institute in America is one of many think tanks funded by fossil-fuel-owning families or companies. They often bring in experts to discuss climate denial tactics. You can see these videos online. They have been brilliant at creating 'cultural norms' and memes around the topic. Much of their work is often repeated in online debates. You might be aware of such classics as: 'The science isn't settled', 'The climate has always been changing', 'It's the sun', 'Climate change is an excuse to implement a communist world government' or 'Those whacky climate alarmists just want to stop growth'. See the terrific skepticalscience.com (run by Australian scientist John Cook) if you want to see the real answers to the above claims.

4.

CAST DOUBT ON THE EVIDENCE.

As mentioned earlier, there is a 97 per cent consensus among climate scientists that global warming is real and that we are causing it. A leading climate scientist recently told me it is likely to be stronger now as the consensus study was published in 2016. That's as strong as the consensus that smoking causes cancer – and yet the denial campaign has managed to soften this claim. Climate deniers constantly attempt to debunk the figure, often by citing a petition of 31,000 scientists that claims global warming is neither happening nor a result of human activity. It's called The Petition Project. What they don't tell you is that just 12 per cent of the signatories have a degree qualification in earth, atmospheric or environmental science. In fact, you don't even have to be a practising scientist to sign the document. It's like asking an astronomer to comment on the metabolic pathways of fructose in the liver.

Take note of how the denial movement has also changed its tune. As nature flexes its muscles and hammers home the truth, the movement has shifted from a position of 'it's not real' to 'it's real but not our fault' to more recently 'it's real but may be great for humanity'.

5.

ATTACK SCIENTISTS.

Many climate scientists report regular attacks via email and social media, as well as direct targeting of their integrity and funding sources. Some of these are from people who simply believe that they are correct and the scientists are wrong; others are deliberate and calculated.

6.

FAKE EXPERTS.

These guys are often wheeled out to give the impression that there is still debate. The thinking behind it goes: 'The science isn't settled so we don't need to make any rash decisions.' (This worked for tobacco.) Giving air time to fake experts is called 'magnifying the minority'. The media can be unhelpful here, too. A great example of this was the most recent Intergovernmental Panel on Climate Change (IPCC) report in 2018 urging us, yet again, to act. This report involved 91 climate scientists (the majority volunteering their time) and 6000 papers (to go with the now more than one billion data points around the globe to measure the changing climate). Many climate-change-denying media organisations used a single report from one 'expert' and held this up as a counterargument.

7.

SPREAD MISINFORMATION.

Apart from the deliberately misleading websites and front groups I discovered ('The Centre for the Study of Carbon Dioxide and Global Change' and 'Plants Need CO2' were my favourites), it turns out that news channels are also very good at spreading misinformation and incorrect climate science. The Union of Concerned Scientists found that 70 per cent of the climate science reporting on Fox News was incorrect and 30 per cent of the science on CNN was wrong.

When the US political consultant Frank Luntz was asked how to prevent public support for climate-mitigating policies he replied: 'Make the lack of scientific certainty a primary issue.'

DRAW
AND SE

DOWN
UESTER

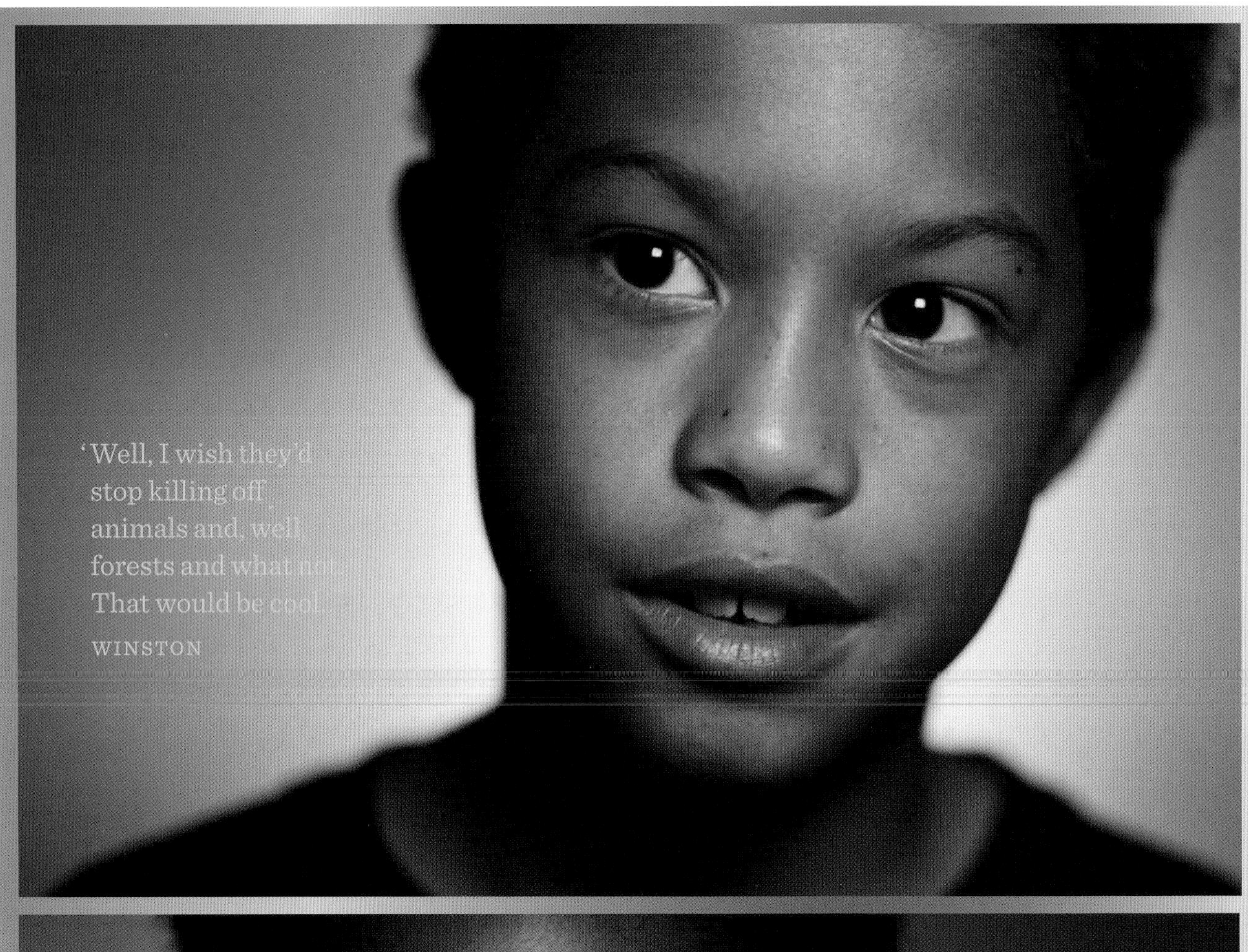
'Well, I wish they'd stop killing off animals and, well, forests and what not. That would be cool.'
WINSTON

'Be more healthy. That is all I really have to say about the food.'
MILES

Climate change denial has not only delayed us from taking action on lowering our emissions, but it has also prevented wider discussion of a crucial aspect of our dilemma. Many people believe that if we just stop burning fossil fuels we'll be fine. If we switch to renewable energy, all buy Teslas and perhaps skip the odd burger, then we'll fix global warming. Unfortunately it isn't that simple. Even if all emissions stopped right now, global warming would still continue for centuries due to the billions of tonnes of heat-trapping carbon we have already poured into our planet's delicate atmosphere. We need to remove it – urgently.

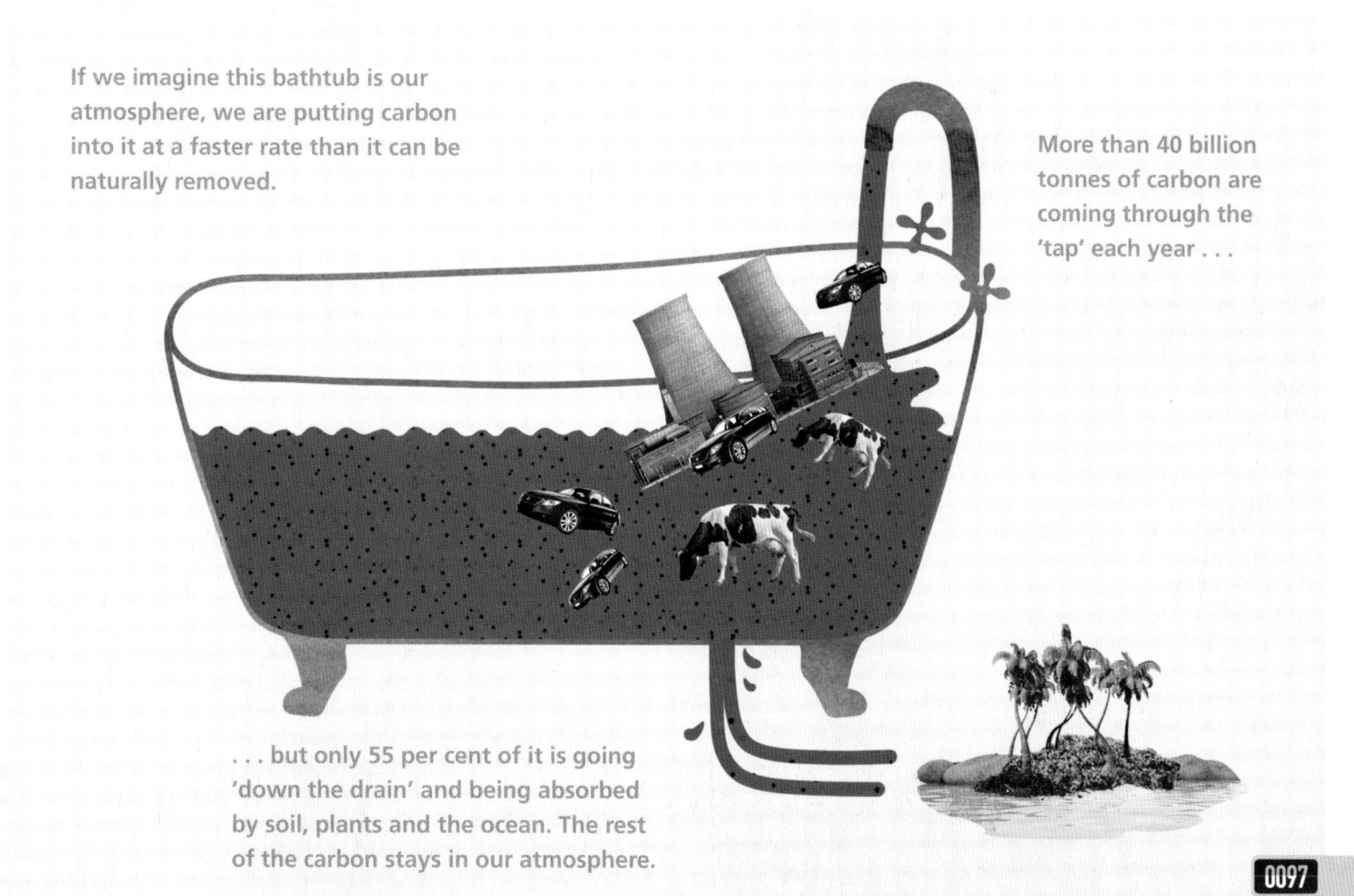

If we imagine this bathtub is our atmosphere, we are putting carbon into it at a faster rate than it can be naturally removed.
More than 40 billion tonnes of carbon are coming through the 'tap' each year . . .
. . . but only 55 per cent of it is going 'down the drain' and being absorbed by soil, plants and the ocean. The rest of the carbon stays in our atmosphere.

'If we cut emissions to zero today, we'd still be in trouble. We've already committed ourselves to changes like sea level rise. In order to mitigate climate change we need to do two things: we need to cease our emissions but also remove carbon from the atmosphere and store it. One of the best places to do that is in soil and biomass.'

ERIC TOENSMEIER,
THE CARBON FARMING SOLUTION

WE CAN DO IT

Carbon has copped some bad press of late, but it is actually the miraculous building block of all life on Earth. It should be celebrated daily. For billions of years, carbon has cycled between the ocean, fossil fuels (often buried underground), the atmosphere, the soil and living organisms.

The opposite picture of Earth would have looked a little different 200 years ago. Our burning of fossil fuels and intense farming practices have moved carbon from the ground (fossil fuels and soil) to the atmosphere and ocean, where it is now doing us harm.

Learning about the various natural solutions that help move carbon from the atmosphere to store it in soils or plants was perhaps the most exciting part of my journey. The following solutions are simple, biological by nature, and have beautiful cascading benefits for people, animals and our environment.

The Carbon Cycle

CARBON AS FOSSIL FUELS
14.9 per cent
(and shrinking)

CARBON IN ATMOSPHERE
1.5 per cent
(and rising)

CARBON IN THE OCEAN
77.4 per cent
(and rising)

CARBON IN LIVING ORGANISMS
(plants, animals, soil)
6.2 per cent
(and shrinking)

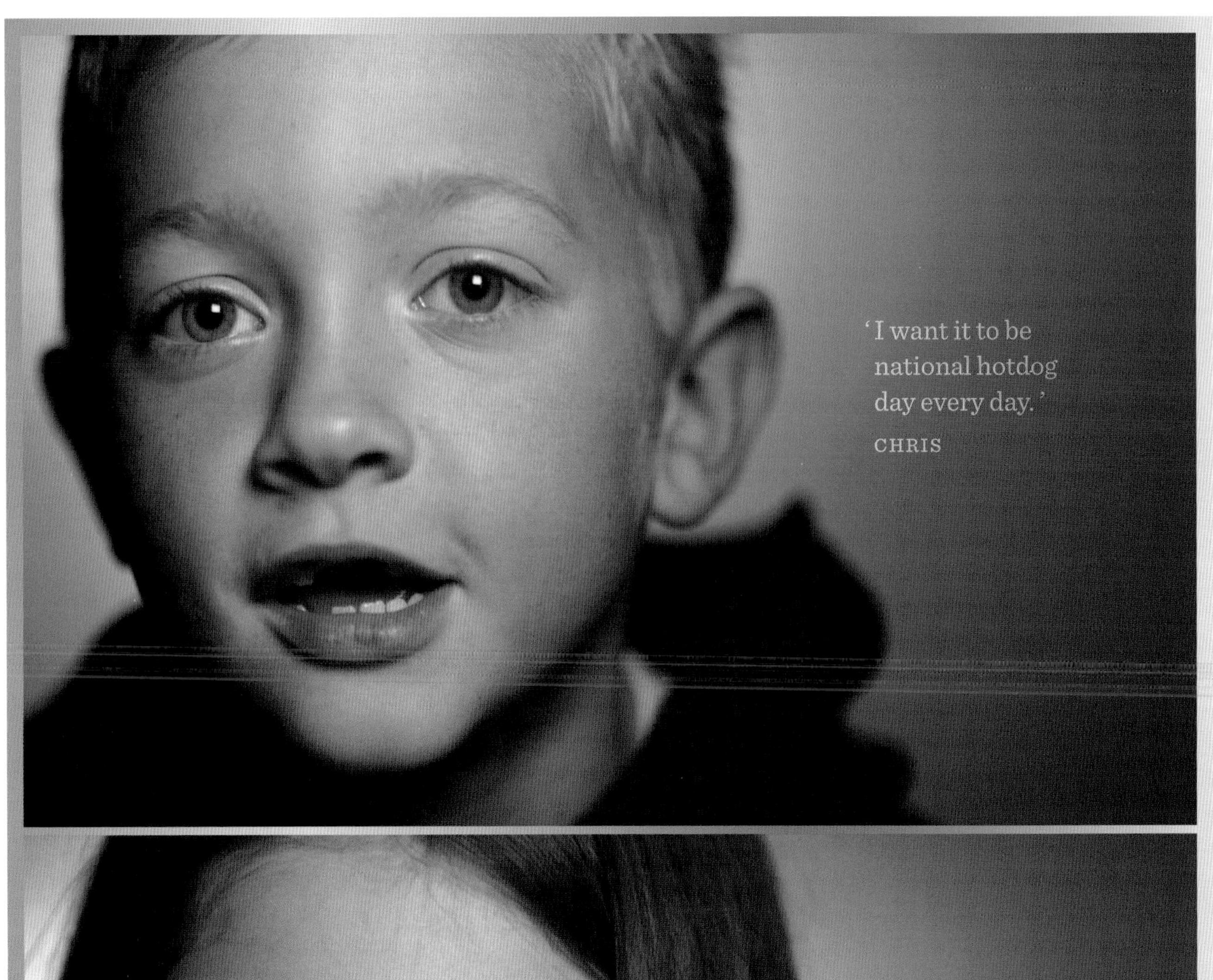
'I want it to be
national hotdog
day every day.'
CHRIS

I do like bacon. Bacon is
nice. But I still like pigs
and I have a pig toy and
when I eat bacon I feel
kind of sad.'
TESSA

REGENERATIVE AGRICULTURE

In a San Francisco green-screen studio that eventually became the top of a wind turbine, Paul Hawken from Project Drawdown shared one of the biggest surprises from their modelling. When the 100 most substantive solutions for reversing global warming emerged, eight of the top 20 solutions were food-related, while the energy sector had just five in the top 20.

'If the impact of a sector is great,' Paul told me, 'then the opportunity is also great. When you change agricultural practices related to food, you can do two things. One is you stop emitting carbon, but you're also sequestering carbon. So you're flipping a whole sector. Not only do you stop putting it up but you're also bringing the carbon back home. That's why it's the number one sector.'

'People ask me, what can we say to farmers? The answer is, we can't do it without you. We can't mitigate climate change without agriculture so we have to find ways for agriculture to become part of the solution, instead of part of the problem. It turns out there are many, many hundreds of such ways.'

ERIC TOENSMEIER,
AUTHOR AND PROJECT DRAWDOWN
LEAD RESEARCHER

To help us clarify the solutions, it's important to first understand how current agricultural practices are hurting our environment:

» Land clearing and deforestation for agriculture makes up 15–18 per cent of our food system's greenhouse gas emissions, with the vast majority (70–90 per cent) of all land clearing being for sugar cane, soy and maize plantations. (Most of this soy and maize is grown to feed livestock, not humans.) Since 1850, the clearing of land plus constant ploughing or tilling of the soil has released 155 billion tonnes of carbon into the atmosphere.

» The livestock sector contributes 14.5 per cent of total global emissions, with 39 per cent of that figure attributable to methane (which is predominantly from cow burps, not farts as most people think, plus manure).

» Our overuse of chemicals such as nitrogen and phosphate has not only destroyed the soil's ecosystem, but also polluted rivers and the ocean as these chemicals are washed into waterways.

» A combination of studies suggest that somewhere between 5 and 7 billion animals (including mice, reptiles, wild birds, freshwater fish and orang utans for palm oil) are killed each year during land clearing and harvesting for plant agriculture alone.

'When the first settlers arrived in Iowa, the soil was 16 feet (4.8 metres) deep, today it's 12 inches (30 cm). We have basically been mining carbon from the soil.'

PAUL HAWKEN

1

'Plants use carbon dioxide and energy from the sun to create simple sugars. The plant uses some of these sugars to grow; the rest of the sugars are pumped into the soil through the roots.'

With these confronting facts in mind, I headed for the town of Shepparton in Victoria. It is the home of the Yorta Yorta people, it houses Australia's largest supplier of canned fruits and soups, and its roadhouse makes a pretty decent egg and lettuce sandwich.

I'd arranged to meet a legend of Australian farming, Colin Seis. Col lost his farm, his house, 20 kilometres of fencing and 3000 sheep to a bushfire in the 1970s. Late one evening, when he and a mate were 10 beers deep at the local bar, probably listening to the newly released Rod Stewart single 'Da Ya Think I'm Sexy?', their conversation turned to the pitfalls of modern farming. Despite a sore head (and eardrums), their discussion still seemed valid the next morning, so Col began experimenting on his now blank canvas of a farm with what is referred to today as 'pasture cropping'.

Over the past 35 years, Col has honed pasture cropping and other agricultural techniques that he now teaches farmers around the world. All of these techniques improve the soil's ecosystem while producing equal, or in many cases higher yield, and minimising weeds and chemical use.

I asked Col his secret and his answer was simply: 'Plants, plants and more plants!' Col tells farmers that the way to improve their soil is to use plants to pull carbon out of the atmosphere where it is doing us harm and put it into the soil where it can do us wonders. I made Col strike a 'Mr January' pose from an Aussie farming calendar while he explained how it works:

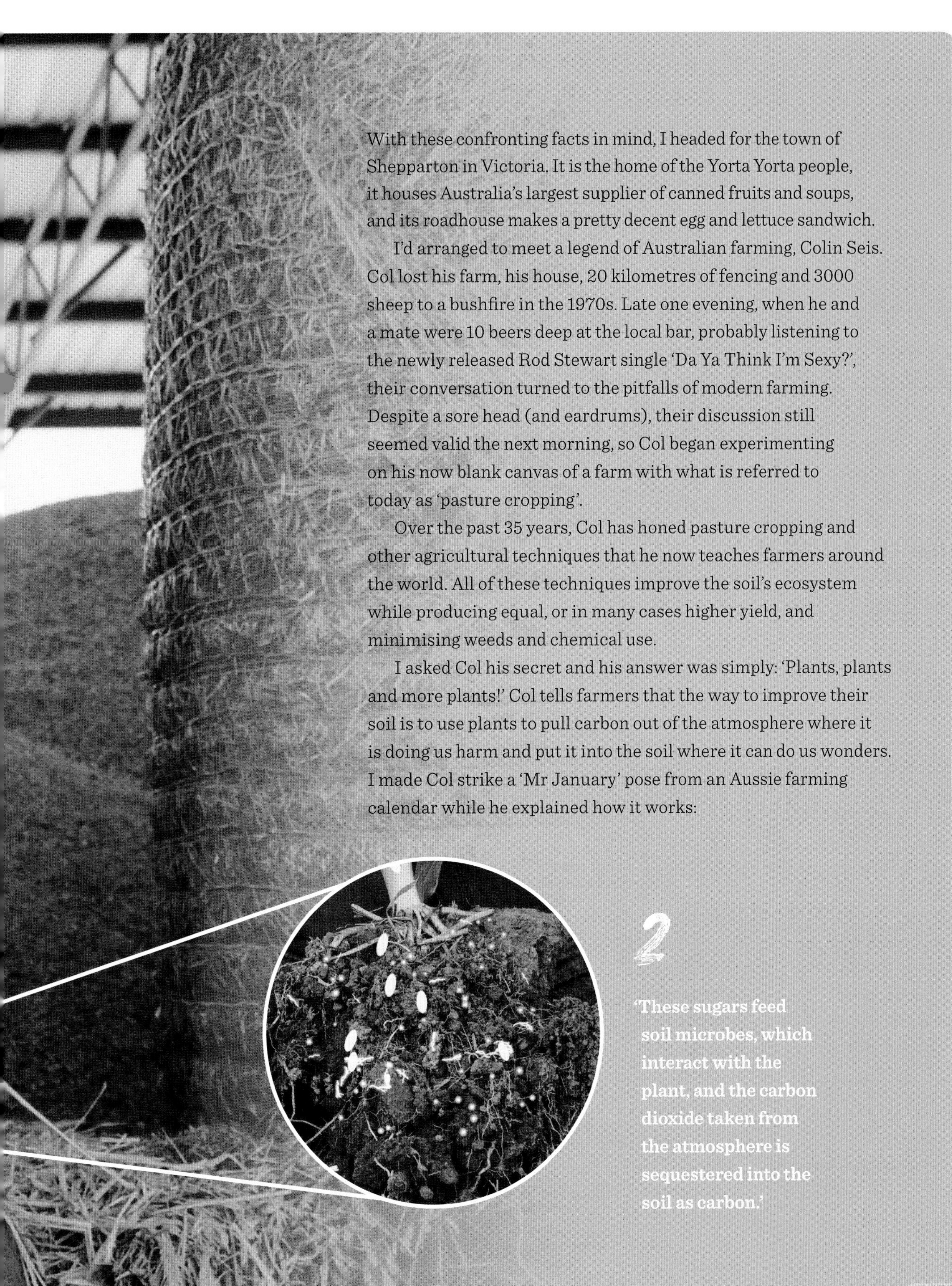

2

'These sugars feed soil microbes, which interact with the plant, and the carbon dioxide taken from the atmosphere is sequestered into the soil as carbon.'

The standard planting of annual grain crops, like rye, oats or wheat, involves first tilling or ploughing the soil, adding fertiliser, then sowing seeds or planting seedlings. Col's pasture-cropping technique was very different. It involved planting rye, oats or wheat amongst the native grasses that already existed on his land. This means he didn't dig up the land first (and release carbon), he simply scattered the seeds amongst the grasses and allowed the crop to rise above them. He then jumped into his harvester and scooped the crop off the top.

Col discovered that he could then bring in his livestock (sheep) to eat the native grasses after the crops had been harvested. He'd let the sheep chomp the grass to a certain level, then move them to another area to allow the grasses to grow back quickly and pull more carbon into the soil. The livestock would then be brought back at a later time to repeat the process. This practice mimics the grazing patterns of herbivores throughout history, which were constantly moved across the grasslands by predators. They would rarely stay in one location (as animals are often kept today) but would chomp the grasses, get chased away by predators, the grasses would grow back and the cycle would continue when the next herd came along (this explains why the soil was so deep and rich under prairies and grasslands). This is how our landscapes evolved, with animals playing a crucial role in soil health.

By planting his crops into the pasture and then bringing in his sheep after harvest, Col has cleverly managed two practices on the same piece of land for a whole year. Most importantly, the soil ecosystem remains intact, requiring fewer chemicals, which produces healthier food without releasing carbon into the atmosphere.

Col told me that soil is an 'unknown universe' and that a spoonful of healthy soil can contain around 6 billion microbes. ('Who the f#*k counted them?' I hear you say.) This soil ecosystem needs to be lavished with praise. It provides plants with nutrients (more nutrients = healthier food), it casually helps reverse global warming by storing carbon, and, while juggling these civilisation-saving traits, it also plays a pivotal role in buffering us from the short-term climate pain that awaits us. Col told me that by raising the carbon content of the top 30 centimetres of soil by just 1 per cent, the water-holding capacity of that soil increases by 166,000 litres per hectare every time it rains. This means that, as droughts ramp up, our soils will be much more protected. And thus, so will our food supply. It also means that when a big rain does dump down (warmer air holds more moisture, so this will happen), the water won't just wash off the top like it does with concrete and be wasted. The precious water will be captured and stored by a more 'spongy' soil. This is a point we need every world leader and decision-maker to understand, and it is why we should begin paying farmers to put carbon back into their soils. The United Nations estimates that we only have 60 years of topsoil left with our current methods. That's just 60 more harvests. The stakes couldn't be any higher to implement a change.

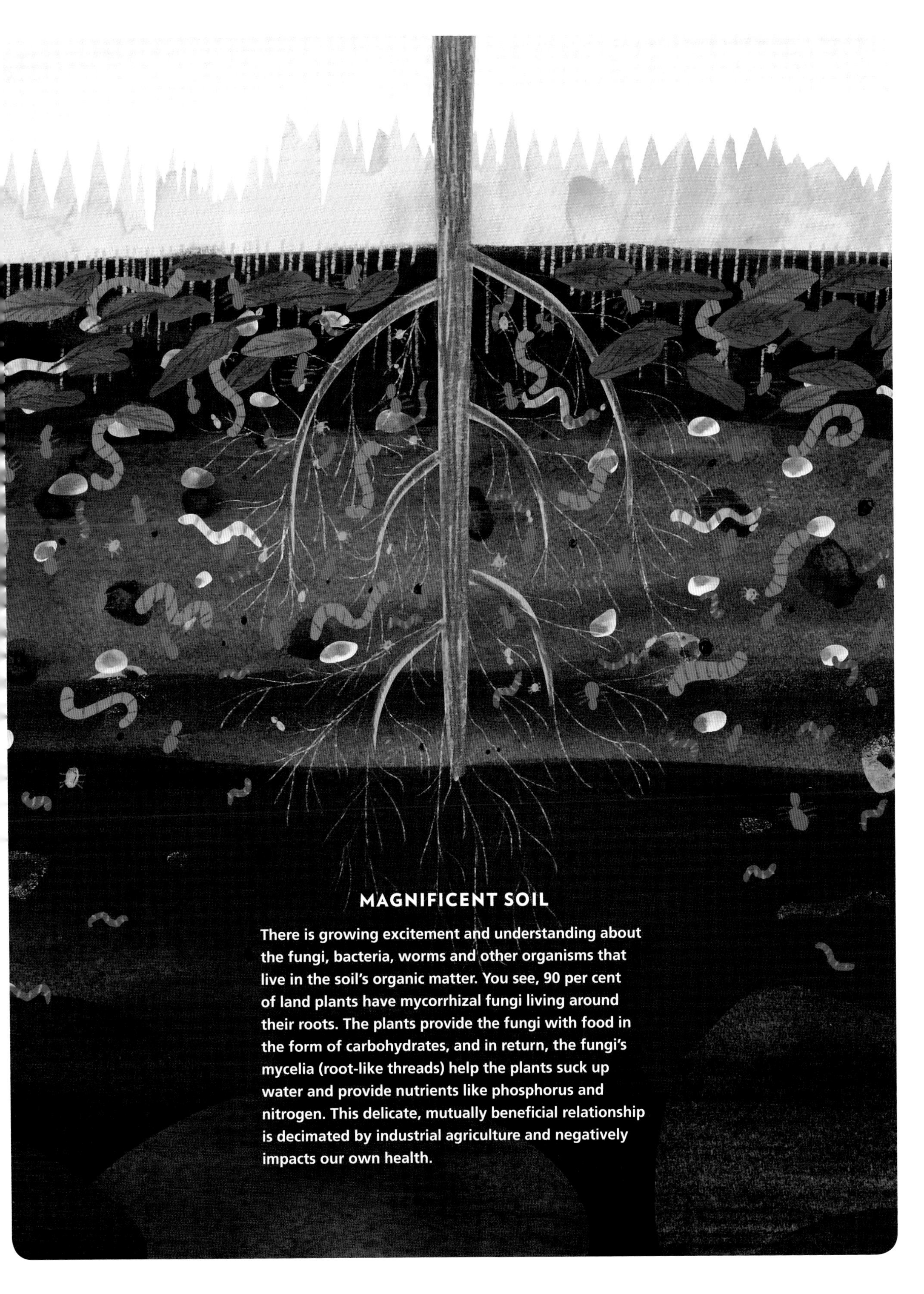

MAGNIFICENT SOIL

There is growing excitement and understanding about the fungi, bacteria, worms and other organisms that live in the soil's organic matter. You see, 90 per cent of land plants have mycorrhizal fungi living around their roots. The plants provide the fungi with food in the form of carbohydrates, and in return, the fungi's mycelia (root-like threads) help the plants suck up water and provide nutrients like phosphorus and nitrogen. This delicate, mutually beneficial relationship is decimated by industrial agriculture and negatively impacts our own health.

What Col and many other planet-saving farmers are doing is described as 'regenerative agriculture'. These are the practices that can restore land damaged by excess tilling or chemical use.

I have spent time on a few regenerative farms and, although they were in different countries and different climates, they all shared the same vitality and abundance. Two of the farms – one in Rodale, Pennsylvania and the other in Dubbo, New South Wales – were straight out of a Disney film. (I can guarantee you won't read 'Dubbo' and 'Disney' in the same sentence very often.)

I remember the farms I visited as a child being dry, with hard, compacted soil and not many signs of life. But as Col explained to me, once you grow the plants and put more carbon and organic matter back into the soil, not to mention rotate some livestock that deposit their 'natural fertiliser' all over the place, the Fairy Godmother waves her wand and the magic begins to happen. First come the worms and microbes to break down dead roots and plant matter in the soil, and the flies and beetles to feast on the dung, then the web-spinning spiders rock up to catch the flies and a host of other insects join the party, which attract the birds that now have delicious tucker. Then come the other animals such as kangaroos that graze on the native perennial grasses, or perhaps you'll spot a platypus luxuriously bathing in the now clean, chemical-free, spongy-soil filtered water. Disney, I tell you – in Dubbo!

Healthy soil with carbon (left) holds more water; unlike the chemical-ridden soil (right), it doesn't run off the top.

In Rodale, I even had butterflies landing on me as I sampled carriage-size pumpkins, succulent tomatoes and other vegetables. How's that for an uncle Walt cliché? Our film's cameraman thought he had gone to heaven, but I suspect this was also due to the workers at Rodale being predominantly radiant, organic-kombucha-drinking young women.

The handful of regenerative farms I have been to in Australia all had neighbours who still practised industrial methods. The differences were stark, and within just metres of the fence line. Of course, many of those farmers have been farming this way for generations, with good results, and see no reason to change. And before I get too carried away, let me apply the utopian handbrake once again by pointing out that switching to a regenerative model can be tricky, unpredictable and potentially a little frightening. It can take time and money to produce yields. With farmers already under the pump in this country, any reticence to change is completely understandable. This is why financing transitions to these practices is so important. A common-sense approach would see carbon polluters pay a penalty, and that money then used to pay farmers to return the carbon to the soil. But if you thought 'Dubbo' and 'Disney' in a sentence was rare, try 'common sense' and 'politics'.

When we understand the impacts of both industrial and regenerative farming, we realise how important our food choices become. As we will see in the next few pages, every decision we make with food can have a positive or negative impact on our environment, and this is why many believe that a regenerative 'real food' movement is one of the great hopes we have for reversing global warming. It is leaderless, has a potential army of billions, and will improve the health of its participants and the planet. I hope that in the next few years soil health (and its critical relationship with the health of our food, water and environment), receives the same mainstream attention that energy has.

'We've all grown up with pictures of farms with rows and rows of food going into infinity. This kind of farming absolutely destroys the soil. It tills it – it breaks it up twice a year, which emits carbon. It uses lots of chemicals – mineral fertilisers and herbicides like Roundup – which destroy the biota in the soil itself. Regenerative Agriculture stops this process and takes us back to the cycles that have created the abundance and beauty of the Earth itself.'

PAUL HAWKEN

THE *MEAT* CONTROVERSY

Apart from Kraft's 2009 attempt to sell us a 'Vegemite cream cheese' spread, it's hard to imagine a more controversial food topic than meat. Like discussions of politics and religion, there often seems only to be room for polarised thought. When it comes to meat, you are either a cold-blooded, climate-destroying meat-eater or a virtue-signalling, animal-worshipping vegan. It's either Team Paleo or Team *Cowspiracy*.

These divided camps do wonders for our need to belong ('my tribe'). They also provide a sense of being on the 'right' side of something, along with a frisson of superiority. But the truth is far more nuanced than a simple black or white proposition, and nuance is difficult when you are conveying an idea in just 140 characters.

The following is not a pledge of allegiance to either camp, but an attempt to bridge the gap between the two – to share what a fast-flowing internet brimming with clickbait may not often do. It is what Eric Toensmeier, author of *The Carbon Farming Solution* and head researcher for Project Drawdown, brilliantly calls 'teaching the controversy'.

The first thing to mention is that we are in a very fortunate position in wealthier nations to even have this discussion. Many people around the world rely on cattle or goats not only for their income but also for their calories, as their particular region is unsuitable for growing the types of crops we often take for granted in our supermarkets.

What I have explored in the film and now this book are the environmental impacts of meat consumption. The issue of animals being sentient beings and whether or not we should eat them is an entirely different and valid discussion. It is an intensely subjective topic that I grapple with daily in my own life. The nuance here is that no matter what foods you eat in our current agricultural system, there will be loss of life on a large scale. As mentioned earlier, plant agriculture alone is responsible for the death of at least 4 billion creatures per year, including mice, reptiles, wild birds, freshwater fish etc.

Environmentally, the damage from livestock production (meat) comes in different ways. As well as the land cleared to raise cattle (70–80 per cent of Amazon deforestation is for cattle ranching) and the impacts of using one-third of the world's cropland to grow food for animals (think of all the constantly tilled soil releasing carbon into the atmosphere, plus the huge amount of chemicals used on the land, destroying the soil's ecosystem and washing into nearby rivers and waterways), there's also the issue of methane emissions. Methane is a greenhouse gas 21 times stronger than carbon dioxide, although it only hangs in the atmosphere for 9–15 years. It is produced from natural sources like wetlands but also when plants decompose in an anaerobic environment (i.e. without oxygen), such as when buried in landfill, or when fermented in the guts of livestock such as goats, sheep and cattle.

Methane (from all sources) makes up 16 per cent of total greenhouse gas emissions, with methane from livestock making up around 44 per cent of that figure. (Unfortunately, despite its best intentions, the documentary *Cowspiracy* incorrectly reported that 51 per cent of total greenhouse gas emissions came from animal agriculture.)

Despite the misconceptions, and the fact that there are some livestock-raising techniques that can offset the methane emissions, the reality is that we are going to need to reduce our meat consumption. This doesn't necessarily mean becoming vegan or vegetarian, unless you choose to be, but after extensive research and consultation, the Project Drawdown team recommends people in wealthier nations lower their meat consumption from 110 grams per day to 50–60 grams a day.

But again, there is nuance. The type of meat we eat is critically important.

Whether you support Team Paleo or Team *Cowspiracy*, I'm sure you'd agree that feedlots are not a humane option. Animals often endure barbaric conditions; they are unable to exercise and fed on grains or other unsuitable foods that leave them in poor health. (They are sometimes given other random rubbish like Skittles. Yes, Skittles. Google: 'truck crash skittles cattle'.) Grains are totally unnatural for ruminants to eat and can damage their gut health. As a result, antibiotics and growth hormones are often added. This 'hot bed of good health' is then passed on to the humans who eat the meat.

The issue of methane and feedlots is complicated, as studies have shown that the animals emit less methane in feedlots than on lush green pastures. This is because the grain requires their insides to do less work (there's less fibre) and so less methane is released. However, the methane emitted from their manure in the confined feedlots is higher than in a lush green field because the poo on the land can be utilised by insects, soil and plants. I hope you are starting to see the subtleties that don't fit within a one-sentence social media headline.

'If you look to the history of livestock, no one was growing grain and feeding it to animals 1000 years ago. That wasn't what happened.'

ERIC TOENSMEIER

Ideally, we would return animals to the land in regenerative agriculture. In fact, this is a recommendation of the Food and Agriculture Organisation of the United Nations (FAO), because animals are crucial to us reversing global warming. The earth co-evolved with animals, so it doesn't make sense to remove them from the equation. Many farmers, including vegan farmers, are using carefully managed and rotated grazing practices (like Col's) to pull more carbon into their soils. They are also using animals to replace fossil-fuel machinery on their farms and, as I just explained, as a source of super-poo natural fertiliser. These are the farmers we would ideally buy our 55 grams a day of meat from in the future.

I have seen technologies on this journey that allow you to track your food and learn what chemicals were or weren't used on them. I have no doubt that in the next few years we will be able to track not only where our meat has been produced, but also what farming practices were employed. This still wouldn't mean we could eat burgers every day, but it would satisfy the needs of meat eaters while actively helping the planet. 'If we were to raise livestock only in ways that are mitigating climate change and feed them on crop residues, food waste or grass,' Eric Toensmeier told me, 'I'm quite confident there would be less to eat than what the average American eats now. But there would be some.'

While there is no doubt that more people in wealthier nations switching to a vegan or vegetarian diet would benefit our environment, the solution is not straightforward. If the grains and legumes are grown in monoculture fields that are regularly tilled (releasing carbon), fertilised, and sprayed with weedicide and/or herbicides (damaging soil health and killing an array of small animals), our environment would still suffer. If, on the other hand, people source their grains and legumes from farms that employ regenerative practices, the impact would be far less. And the increase in demand would mean more and more farmers would be encouraged to produce food in this way.

The reality is that livestock farming produces much less food per hectare than crop farming (at least where rainfall and terrain permits crops to be grown). This is true for both grain- and grass-fed meat. So, to feed a rising population without deforestation and the emissions that come with that, reducing meat in the western diet is essential. Crops are simply a more efficient producer of food per unit area.

'In the very legitimate concern about the methane that's produced by livestock,' said Eric, 'we're overlooking the fact that there are cattle production techniques that sequester more carbon than the methane they emit. Some of them profoundly so.' Now these practices aren't accepted in the mainstream just yet and may invite a bewildered look from the pimple-faced worker at your local supermarket, but 'silvopasture' and 'intensive silvopasture' are terms to get familiar with. Silvopasture is actually Project Drawdown's number 9 solution to reversing global warming, and simply involves livestock being raised on pasture that also has trees. (The term comes from 'silvis', which is Latin for 'woods'.) This combination improves soil health, reduces water use, prevents erosion and provides shade for the animals. According to Eric, silvopasture can sequester 3–10 times more carbon than rotational cattle-grazing alone. (It can sequester around 4.8 tonnes of carbon per hectare per year and has twice the impact of regular livestock grazing on one-fifth of the land. Organic annual cropping and managed grazing both sequester around 1 tonne of carbon into the soil per hectare per year.)

Intensive silvopasture

The 'intensive silvopasture' version sees livestock rotated amongst trees and legume crops that provide food for the animals and fix nitrogen in the soil. Additionally, when the animals are moved on to graze elsewhere, the leaves will grow back rapidly, thus sequestering carbon. A leguminous diet also reduces methane emissions from the livestock and farmers report far greater yields (2–10 times more meat per hectare). Intensive silvopasture was developed in Australia and currently sequesters the most carbon of all the livestock farming practices. (A Colombian study showed intensive silvopasture can sequester 8.8 tonnes of carbon per hectare per year, and up to 26 tonnes when timber trees are incorporated.)

My wife and I don't eat a lot of red meat but when we do we buy biodynamic meats. Biodynamic farms apply regenerative practices with a pinch of the esoteric thrown in. Even if the esoteric parts don't float your boat (burying crystals and whatnot), the care taken of the soil and animals is exactly what the doctor ordered, for both us and the planet. The meats are a little pricey compared to others but the difference in taste is extraordinary. Let's hope that with more demand these prices come down and that all of these terms become more widely used and understood.

'The bottom line or takeaway is that human beings, with animals and the land, can actually create more productive environments, with more biodiversity, more pollinators, better soil, more water retention than we do today. So getting rid of animals is not a pathway to reversing global warming.'

PAUL HAWKEN

Considering the furore of debate around eating less meat, I was pretty shocked to learn that Project Drawdown lists reducing food waste as the number 3 solution to reversing global warming, before eating a plant-rich diet (which sits at number 4).

When you dive deeper this number begins to make a lot of sense. Worldwide we waste around 30 per cent of our food (through production, transport and consumption), and in the US the proportion is 40 per cent. Incredibly, if global food waste was a country, it would be the third biggest greenhouse gas emitter behind China and the USA. 'In the developed countries,' Paul Hawken told me, 'food is wasted mostly on the home and restaurant level, while in the developing world, it's on the farm cold-chain level and getting it to people. Poor people generally don't waste food. Rich people do.'

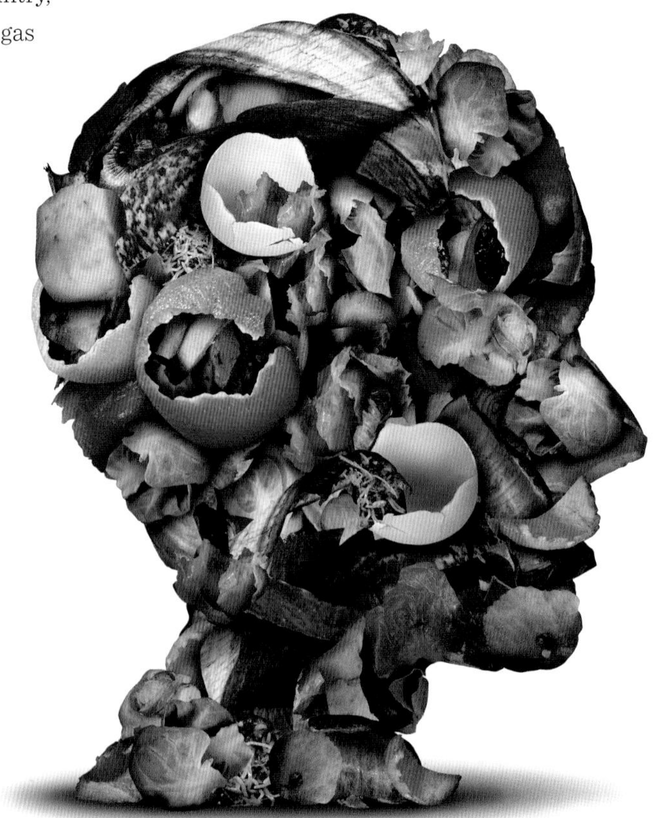

Uneaten food squanders a host of resources, including water, energy, land, seeds, labour, fertiliser and financial investment, plus it generates greenhouse gas emissions at every stage, from production and distribution through to consumption. Even more staggering is that Project Drawdown's findings that make reducing food waste the number 3 solution don't include the methane emissions that occur when food is sent to landfill, as it is too hard to measure. Paul Hawken told me that if landfill methane could be factored in, reducing food waste could be the number 1 solution to reversing global warming.

The good news is that reducing food waste is something we can all take part in right now. By simply finishing your next meal, and thus not sending any food to landfill, you are actively contributing to the regeneration of the earth. I like to call it 'eating for the planet', especially when it's an extra piece of something delicious. 'I mean it's not like ,"How do we do that?"' said Paul Hawken. 'You do it at home, at the restaurant, wherever you are. You can start thinking and acting in such a way that you absolutely eat everything you buy. If you don't do it, then you'll keep buying too much and throwing it away.' See pages 142–43 for other ways to reduce your food waste.

'According to the US Environmental Protection Agency, landfills account for 34 per cent of all methane emissions in the US, meaning that the sandwich you made and then didn't eat yesterday is increasing your personal – and our collective – carbon footprint.'

SCIENTIFIC AMERICAN

'I think that there should be a lot more inventions that can clean up the oceans and things like that.'

RUBY

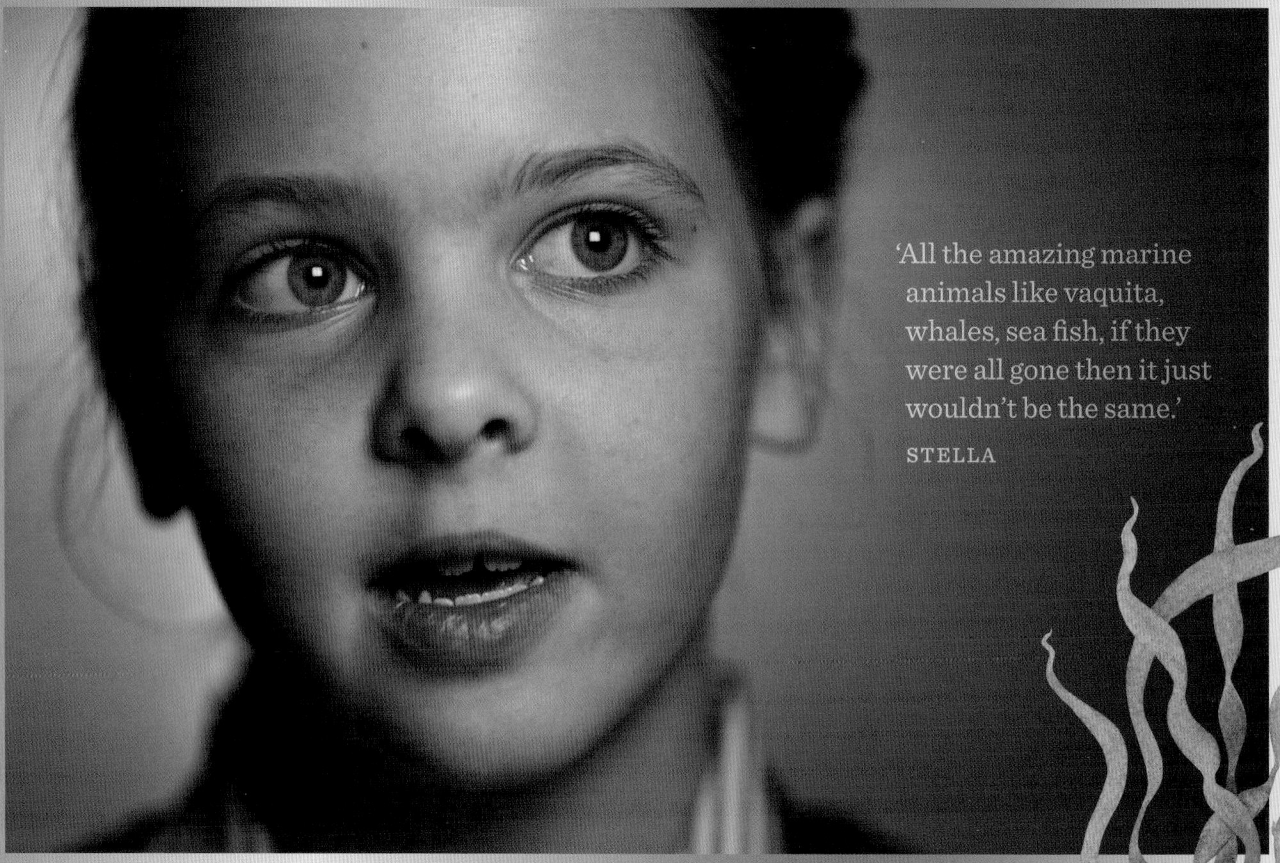

'All the amazing marine animals like vaquita, whales, sea fish, if they were all gone then it just wouldn't be the same.'

STELLA

THE BEAUTIFUL 'SEAQUEEN'

As I left outback Australia with a newfound respect for farmers and the crucial role they can play in reversing global warming, I called Paul Hawken to share my enthusiasm. He very graciously listened and then burst my naive bubble in a few sentences. He told me that regenerative agriculture is not a silver bullet. If we stopped all fossil-fuel emissions tomorrow, our soils and plants could sequester or draw down a huge amount of the carbon already in our atmosphere, but there is a limit to how *much* carbon they can hold. There is a saturation point. And given we won't stop our fossil-fuel emissions tomorrow and are still putting more than 40 billion tonnes of carbon into the atmosphere every year, we are going to need more than regenerative farming to pull that carbon out and avoid centuries of warming.

Paul suggested I pack my bags, book a ticket on an airline with a good lost-luggage record, and head to the Woods Hole Oceanographic Institute in Massachusetts, USA, which is not only a difficult location to say with a mouthful of Milo but is also home to a serene statue of Rachel Carson gazing out to sea. For those unfamiliar with her name, Rachel is famous for kickstarting the environmental movement and her book *Silent Spring* exposed the use of pesticides on our foods and led to the banning of the damaging chemical DDT. (I have a feeling that the scientist I met in Woods Hole may one day warrant a serene statue of his own.)

It was in this idyllic, slightly *Pleasantville* town that I learnt about the simplest, cleanest, least political and most inspiring solution of my whole adventure. If I told you that this one solution could potentially draw down staggering amounts of carbon, restore marine ecosystems, provide enough seafood (brimming with healthy omega-3 fatty acids) for 10 billion people, remove carbon dioxide from the waters to make them more alkaline and chemicals such as nitrogen from ocean 'dead zones', cool the waters over coral reefs to help prevent bleaching, plus provide a material to make plastics and clothing, feed cattle to dramatically reduce methane emissions *and* provide biofuel for energy, you'd probably say, 'This sounds like some Silicon Valley, Elon Musk-funded, Richard Branson-award-winning miracle.' And I'd say: 'Nah, it's just seaweed.'

It's clear to me that whoever named seaweed had absolutely no understanding of its potential. Nothing with such magnificent properties could have 'weed' in its nomenclature. (I would have called it 'seaqueen'.)

Paul had told me to meet with Dr Brian von Herzen, a man with a welcoming softness acquired only by spending hours underwater immersed in nature. Brian is a physicist who used to fly to Europe every year in his small fire-spotting plane (as most of us do). After a glaciologist

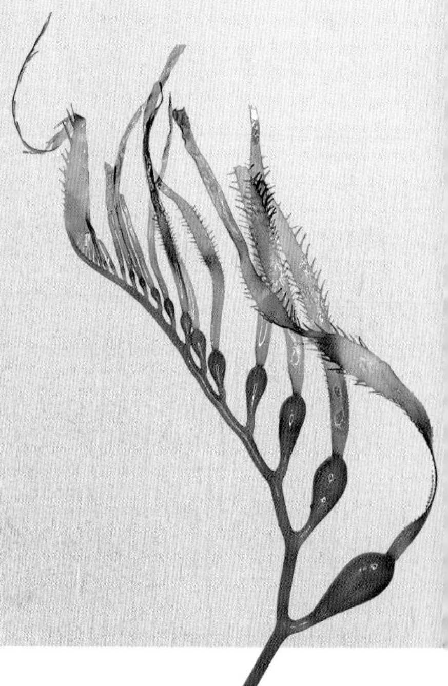

friend suggested he look down at the melting ponds as he flew over Greenland, Brian noticed that hundreds more ponds appeared each year, merging to create huge lakes. He decided to dedicate his whole life to restoring living systems with a focus on the oceans.

One of the first things I discussed with Brian was my own reticence at meddling with the ocean. Haven't we already done enough damage to living systems? Shouldn't we just leave this enormous body of water alone? Again, my naivety was gracefully exposed in a few short sentences. Brian told me that the oceans desperately need our help. 'We have been extracting life from them for centuries, to the point where 90 per cent of the world's fish stocks are over-fished. We have lost 90 per cent of our kelp forests in the past decade off of California alone, and lost huge amounts of kelp north of Perth in the last three years. Plus due to the oceans absorbing so much of our excess carbon dioxide, they are now 30 per cent more acidic than they were 150 years ago.'

Gulp.

And he wasn't even finished.

'But equally concerning is that, with more than 90 per cent of the heat from global warming each year being absorbed by the oceans, the waters are getting too warm. The result is we are losing life in the ocean. Between Australia and the US there are 100 million square kilometres of ocean desert. We need to restore the fish habitats and restore the ecosystems, and seaweed can do that.'

Brian's solution is what he calls 'marine permaculture', where seaweed is grown on frames suspended just below the surface of the ocean. (I suggested it should be called 'seaqueen saviour' but he just stared at me blankly for an uncomfortable few seconds.) He says it 'restores the ocean's overturning circulation. It's as if your leg was asleep and you lost circulation. We have to do that with the oceans. We have to bring the cool, nutrient-laden waters up and restore the overturning circulation.' As Brian spoke, I imagined Mother Nature slumped in a wheelchair outside a hospital with a ciggie in one hand and a sports drink in the other. Type 2 diabetes had kicked in and it was time to take off her leg. I really wanted Brian's solution to work. Here it is:

The tests that Brian and his team have conducted in the South Pacific have been remarkable. They have seen fish and other marine life (even whale sharks!) return to the seaweed platforms. They have also measured the improvements in alkalinity (due to the seaweed drawing down carbon dioxide from the ocean), allowing shellfish and other sea creatures to thrive. (Currently many shellfish and molluscs aren't able to grow their shells properly due to the increased acidity from excess carbon in the oceans.)

Brian explained how the seaweed can drop the temperature of the ocean water it is placed in, which they noticed had an impact on the bleaching of the South Pacific coral reef. This discovery has massive implications for our own Great Barrier Reef and other reefs around the world. Reefs cover less than 1 per cent of the ocean floor, yet house almost a quarter of all marine life.

But amongst all the riches this humble weed has to offer, there are two attributes in particular that could help change the course of history. One is its ability to feed the world. As rising temperatures create havoc with our land crops, these plots could be set up offshore around the world to provide an alternative food source. Not only is seaweed itself full of healthy fats, phytonutrients and antioxidants (and is eaten on the islands of Okinawa whose residents are among the longest-living in the world), but it could also restore thriving fish ecosystems. As Paul Hawken told me on top of the CGI wind turbine: 'You could feed 10 billion people from the protein from marine permaculture alone.' This means we could potentially phase out and replace polluting fish farms with a natural, clean alternative, while at the same time dramatically lowering our meat consumption from those methane-burping cows.

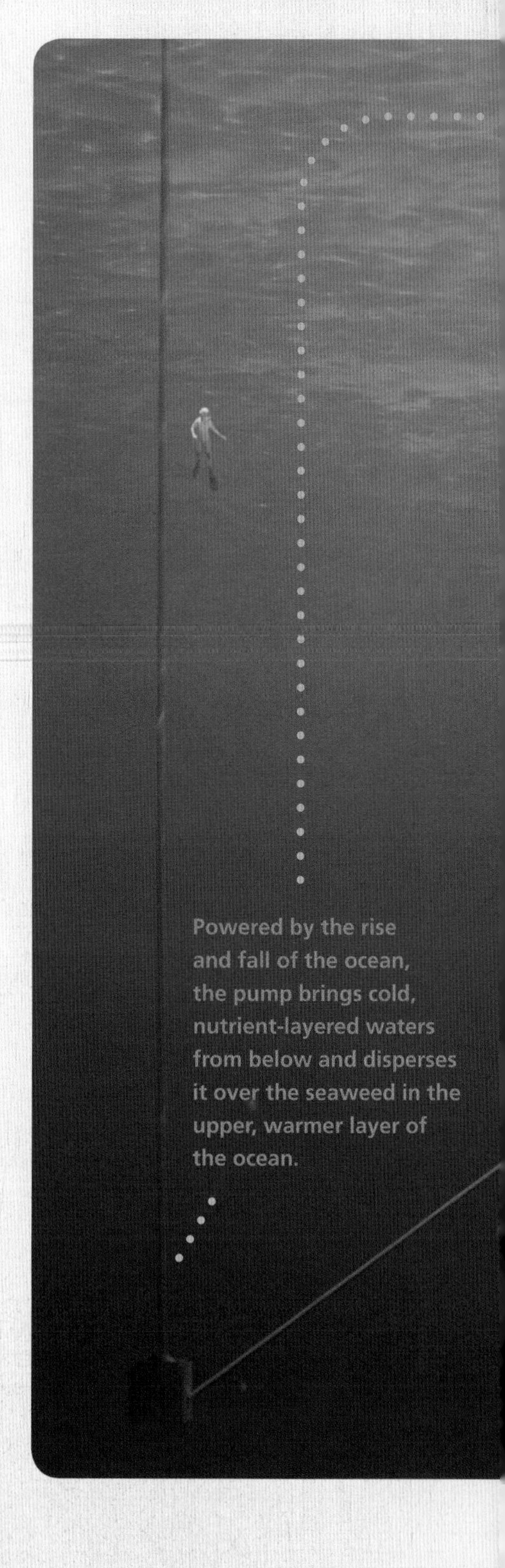

Powered by the rise and fall of the ocean, the pump brings cold, nutrient-layered waters from below and disperses it over the seaweed in the upper, warmer layer of the ocean.

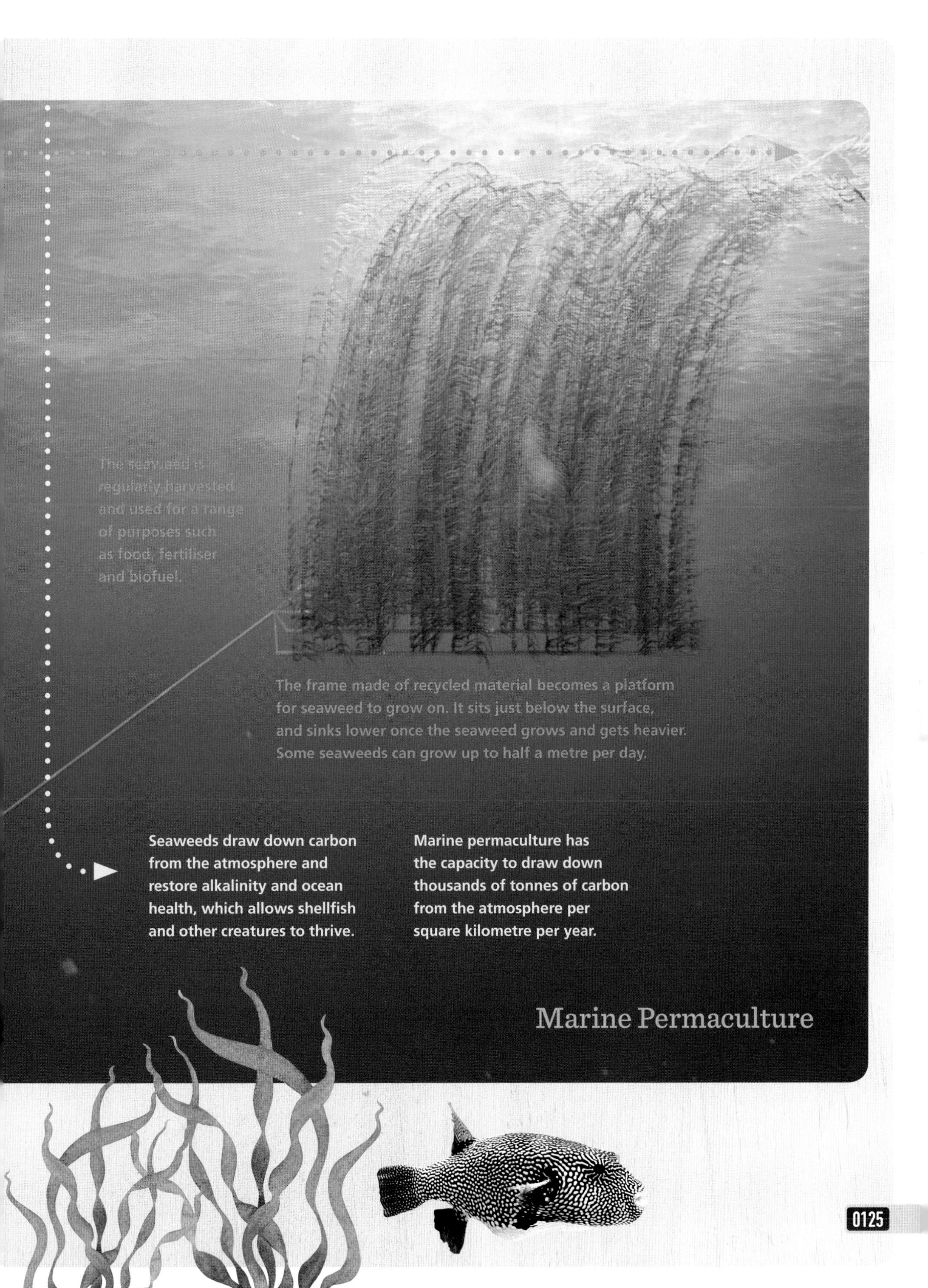

Marine Permaculture

The seaweed is regularly harvested and used for a range of purposes such as food, fertiliser and biofuel.

The frame made of recycled material becomes a platform for seaweed to grow on. It sits just below the surface, and sinks lower once the seaweed grows and gets heavier. Some seaweeds can grow up to half a metre per day.

Seaweeds draw down carbon from the atmosphere and restore alkalinity and ocean health, which allows shellfish and other creatures to thrive.

Marine permaculture has the capacity to draw down thousands of tonnes of carbon from the atmosphere per square kilometre per year.

'Make sure coral looks okay cause it normally looks like a rainbow in the sea with all the fishes. But now it looks like a big, big mess.'

ALIYA

'Instead of having governments that are reacting to disaster, we need governments and communities and businesses that proactively set a new vision that prevents these disasters. What surprises me is the hunger there is for new ideas. I'm struck by the number of politicians, business leaders, city mayors, investors who are clamouring for these new ideas.'

KATE RAWORTH

Two other seaweed specialists, Mark Capron and Jim Stewart from Ocean Foresters, believe some countries may be willing to lend money to climate-vulnerable nations like Bangladesh to enable them to build large-scale seaweed plots, perhaps out of a fear of unprecedented numbers of climate refugees landing on their doorstep. Mark and Jim told me that because of the abundant food the plots would generate, loans could be paid back in a flash. Not to mention the ongoing food supply that would empower local communities by providing jobs and improving health.

But the other history-changing aspect of seaweed is its ability to sequester carbon. It is the fastest-growing plant on the planet (although technically it's not a plant – it's a protist), with some species growing up to half a metre a day and reaching 50 metres in length! This means it is pulling carbon out of the atmosphere at a turbo-charged rate. Estimates vary as to exactly how much carbon, but the numbers are all incredibly impressive (Brian von Herzen suggests around 3000 tonnes of carbon per square kilometre, or 30 tonnes per hectare to compare it with the land). Brian believes the seaweed could be regularly harvested (creating lots and lots of jobs) and used for food, to make fertiliser, fabric, biofuel and, more recently, a plastic substitute that is 100 per cent biodegradable, has a shelf-life of two years and dissolves in warm water. After a period of time (Brian estimates six years) the seaweed can be cut loose from the platforms and, if housed in deep enough ocean, it will sink to the bottom and remain there stored as carbon for a long, long time due to the pressure of the water above.

While this solution is essentially a simple one, it still requires large-scale testing and input from clever engineering minds before it is widely adopted (Brian has recently launched a large test off the coast of Indonesia and in the Philippines). On the plus side, it is currently free of any political baggage and there are no vested interests distorting its purity. Let's hope that this solution can be rapidly adopted by both forward-thinking businesses and governments.

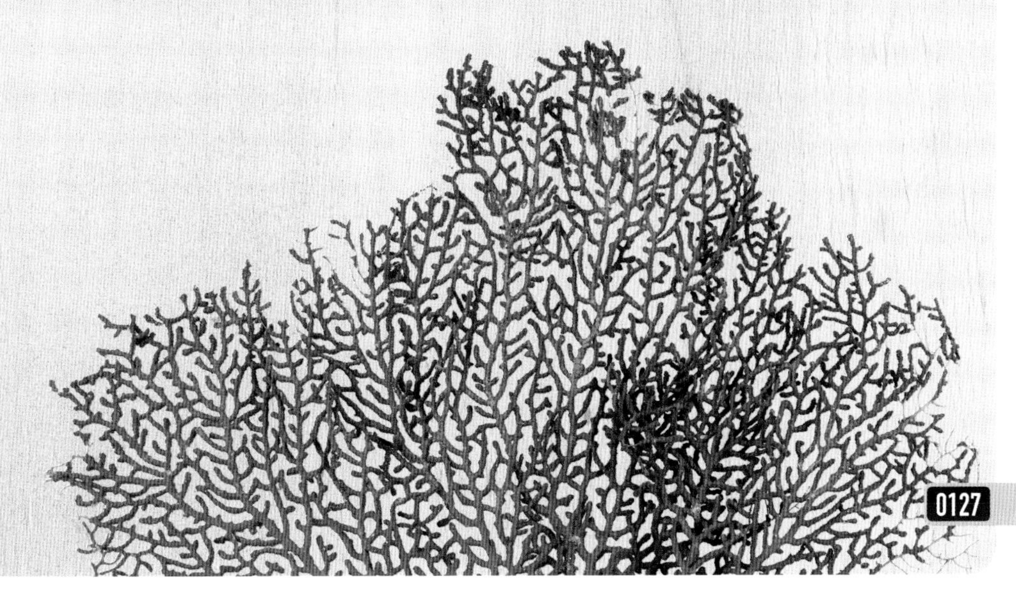

HOW REGENERATIVE AGRICULTURE *& MARINE PERMACULTURE* FATTEN THE DOUGHNUT

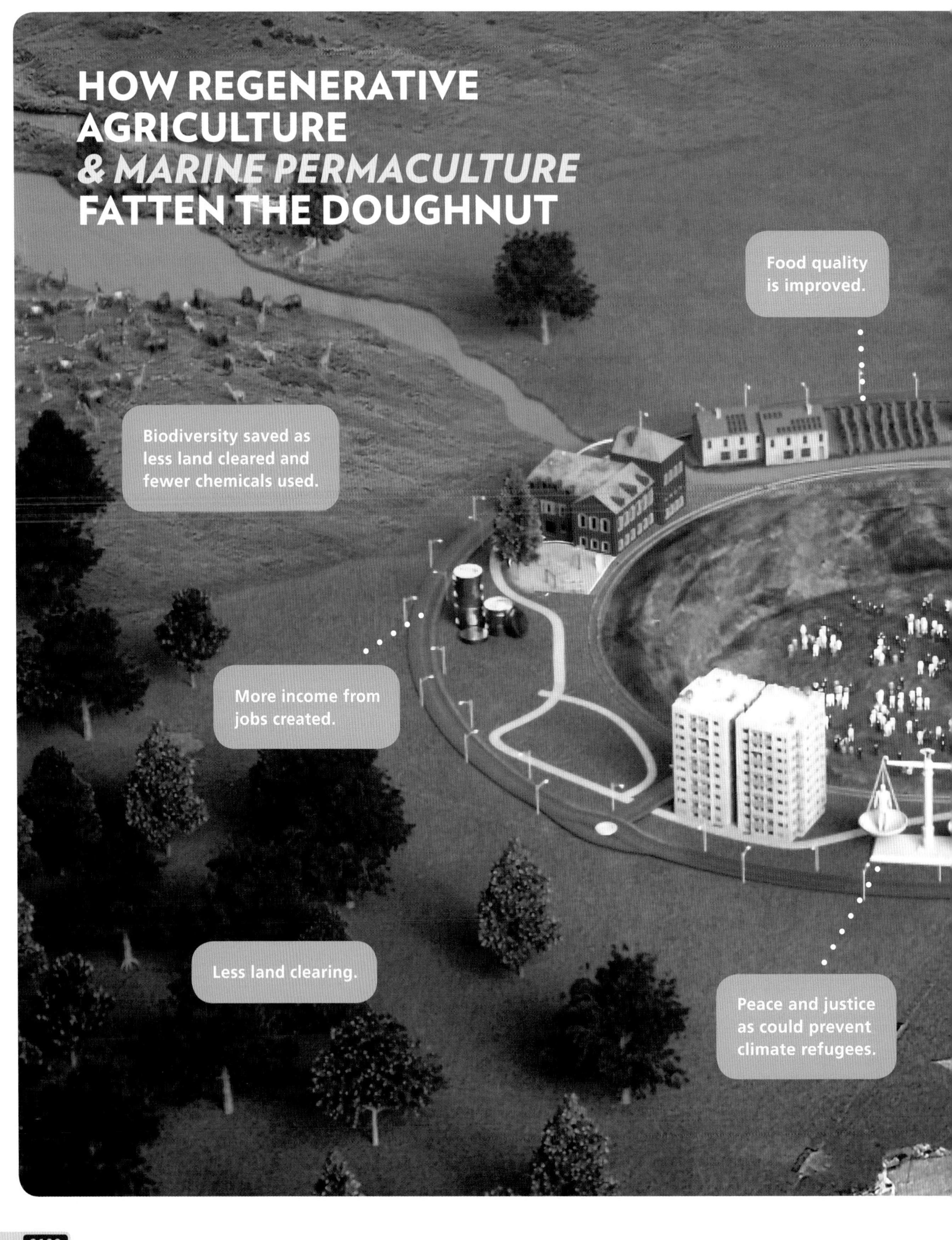

Food quality is improved.
Biodiversity saved as less land cleared and fewer chemicals used.
More income from jobs created.
Less land clearing.
Peace and justice as could prevent climate refugees.

Better health from healthy soil and more seaweed consumption.
More water is retained due to extra carbon in the soil.
Vote
Helps to reverse climate change.
Less acidic oceans.

OTHER WAYS TO *SEQUESTER CARBON* FROM THE ATMOSPHERE

In a time when large high-tech solutions to global warming are being discussed, such as carbon-sucking machines on the edges of our cities or aerosols that reflect sunlight being dispersed into the atmosphere, here are some simpler biological solutions that have cascading benefits for all life on Earth.

Restore tropical forests

Tropical forests once covered 12 per cent of the earth's landmass, but now cover just 5 per cent. Even as deforestation persists, the regrowth of tropical forests can sequester up to a phenomenal 6 billion tonnes of CO_2 a year! That's roughly the same amount of CO_2 as the total annual emissions from the US. According to the World Resources Institute, 'More than 4.8 billion acres worldwide offer opportunities for forest restoration – an area larger than South America.'

Plant more tropical staple trees

Bananas, breadfruit, avocados, brazil nuts, chestnuts, coconuts and carob are all harvested from trees that require no tilling of the soil, need fewer pesticides and fertiliser, and use carbon to grow their tasty goods. They are great for hilly or undulated landscapes that may be difficult for regular crops to grow on. Please note that some forests are being cleared to plant avocado trees. We don't want this: let's leave the forests and plant the avocado trees on degraded lands that need help (Project Drawdown found there are 400 million hectares of abandoned or degraded farmland around the world).

Practise multistrata agroforestry

I witnessed this in Tanzania: coffee, cacao, bananas and macadamias all grown together at different heights on small parcels of land. The soil quality was incredible, as was the taste of the various foods. Why spend all the money on invasive, high-tech fixes to sequester carbon when we can heal the planet while having more chocolate, coffee, bananas and avocados? *Viva la revolucion!*

Plant more bamboo

There are apparently 1500 uses for bamboo (some are pretty random like 'making beer' and 'improving fertility in cows'). But bamboo also happens to take carbon out of the atmosphere faster than almost any other organism ('seaqueen' still reigns). It can grow an inch every hour in spring. It thrives on degraded land and after being cut it sprouts and grows again, thus it's a terrific turbo sequesterer. Bamboo can be an invasive species in some locations, but let's hope we see more bamboo buildings, furniture, cutlery, bikes, tennis racquets, skateboards, beers and pregnant cows in the future.

Practise afforestation

This is the practice of planting carbon-sequestering forests in areas that have been treeless for at least 50 years. Degraded agricultural land is ideal, as are median strips – or how about in some of the empty car parks we'll create by driving fewer cars? This would be supercharged if we used more timber for building – cut the tree, carbon stays stored in the wood in a structure, new tree is planted that sequesters more carbon, and the cycle continues. (Plus, building with concrete is one of the largest single-source contributors to global greenhouse gases: 7 per cent of total emissions.)

Protect peatlands

Also known as bogs or mires, peatlands are basically areas of decomposing plant matter that have developed over hundreds or thousands of years. They make up just 3 per cent of the earth's landmass but are the second biggest storer of carbon after oceans (twice as much as the world's forests!), so preserving them or restoring damaged ones is vital. Eric Toensmeier also told me that peatlands (and wetlands) don't have a saturation point with carbon. They can keep absorbing and absorbing, unlike soils or plants that reach a limit.

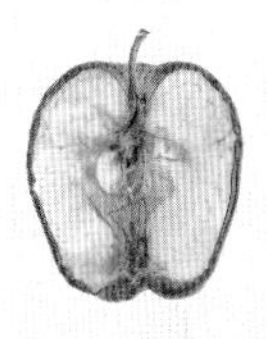

Compost your scraps

Composting food scraps can really improve the carbon-sequestering ability of soil while also preventing more greenhouse gases from entering the atmosphere. As I mentioned earlier, sending our food waste to landfill is a problem due to the methane it emits. Composting can help curb this. Rather than generating methane in a landfill, the composting process converts organic material (your food) into stable soil carbon and makes it available to plants (including the ones in your vegetable garden). Paul Hawken calls it a 'refuse to riches story'. See page 144 to learn how to set up an easy compost system in your yard and contribute to the regeneration.

» Researchers in Queensland have discovered that a particular red seaweed (*Asparagopsis taxiformis* for lovers of Latin) can eliminate most of the methane emitted from cattle or other ruminants when fed to the animals in tiny amounts. (Researchers also found that a different variety of seaweed, dulse, tastes like bacon – perhaps their finest discovery.)

» A certified 'Regenerative Organic' food label has recently been established in the US. It factors in the soil health, animal welfare and social fairness of a product. Check out regenorganic.org for more information.

» France passed a law forbidding supermarkets from trashing unsold food. Instead, supermarkets must donate unsold food to charities or to companies that produce animal feed or make commercial compost. Italy has recently followed suit. Come on Australia, you can do it.

» In Stockholm I witnessed a program where the council collects their residents' food waste, extracts the methane to run their public vehicles and then gives the leftovers to farmers to use as a natural fertiliser to improve their soil. Sweden is *so* 2040.

» The town of Keynsham in the UK now gets 80 per cent of its energy from food waste. The town has built an anaerobic digestion plant and collects residential green waste to power 5000 homes. Another town in Wiltshire, UK, has recently done the same.

» Still on food waste, I met two 23-year-olds in Rotterdam who collect rotten fruit from the town's major market and turn it into leather. They make shoes, bags, wallets and chairs.

- Amazon deforestation rates were reduced by 80 per cent between 2005 and 2015. Brazil is now restoring more than 12 million hectares of its rainforest. (However, at the time of writing, Brazil has a new president who doesn't appear to be a huge fan of forests.)
- Alternative meat options are on the rise, which may help lower demand for meat and therefore reduce the carbon impact of livestock. Some supermarkets are already selling plant based mince. I visited the laboratories of Impossible Foods in San Francisco, who produce meat (I admit 'laboratories' and 'meat' should rarely be used in the same sentence). The 'meat' tasted surprisingly good. It is made predominantly from coconut oil, wheat protein and heme (which is iron extracted from plants and gives meat its 'metallic' flavour). Let's hope the wheat is sourced from farms employing regenerative methods. There are also 'leaf proteins' on the horizon, which sequester carbon as they grow and are very high in protein. They can also be fed to livestock instead of grains.
- Regenerative farming practices are taking off around the world. Paul told me that we're at a crossover point. 'Many farmers doing regenerative farming are not liberals. They know that if they are to continue as farmers, generation after generation, then they have to improve their soil health. They know you cannot make a living by continuing to put more and more chemicals on the soil.'
- A company in Taiwan is the first to use blockchain technology (a highly secure way of verifying information) to track the contents of food. Users of OwlChain only need to scan a sticker on the food. For fruits and vegetables the data includes the date of harvest and any pesticides used. For meats and poultry the data includes the date of birth, antibiotics used (if any), vaccines, and the date of slaughter. In the coming years I can imagine 'soil health' being added and 'farming practice'. This would be a game changer.

WHAT YOU CAN DO TO HELP

Eat for the planet

In a handy turn of events, lowering your factory-farmed meat consumption, plus adding more plant-based proteins (such as peas or beans) and lowering your sugar consumption, is terrific for your own health and equally magnificent for our environment. (Sugar cane is perhaps the most environmentally damaging crop of all time due to land clearing, constant tilling, burning of crop residue and chemical use.)

EAT LESS MEAT

As suggested by Project Drawdown, aiming for 50–60 grams of meat a day is a great start. At home, we have red meat very occasionally and when we do it's either organic and grass-fed or biodynamic (the high cost of these is a great encouragement to eat less!) Look out for a 'Regeneratively Organic' meat label in the future.

This is why we have included a range of delicious plant-based recipes in this book. If enough of us cooked a few of these each week (without creating any food waste), it would make a difference to our children's future.

Greenhouse impact of 15 foods

Please note that these figures are based on conventional agricultural practices. As we discovered earlier, there are ways of raising livestock (or even making cheese) that can actually sequester carbon into the soil.

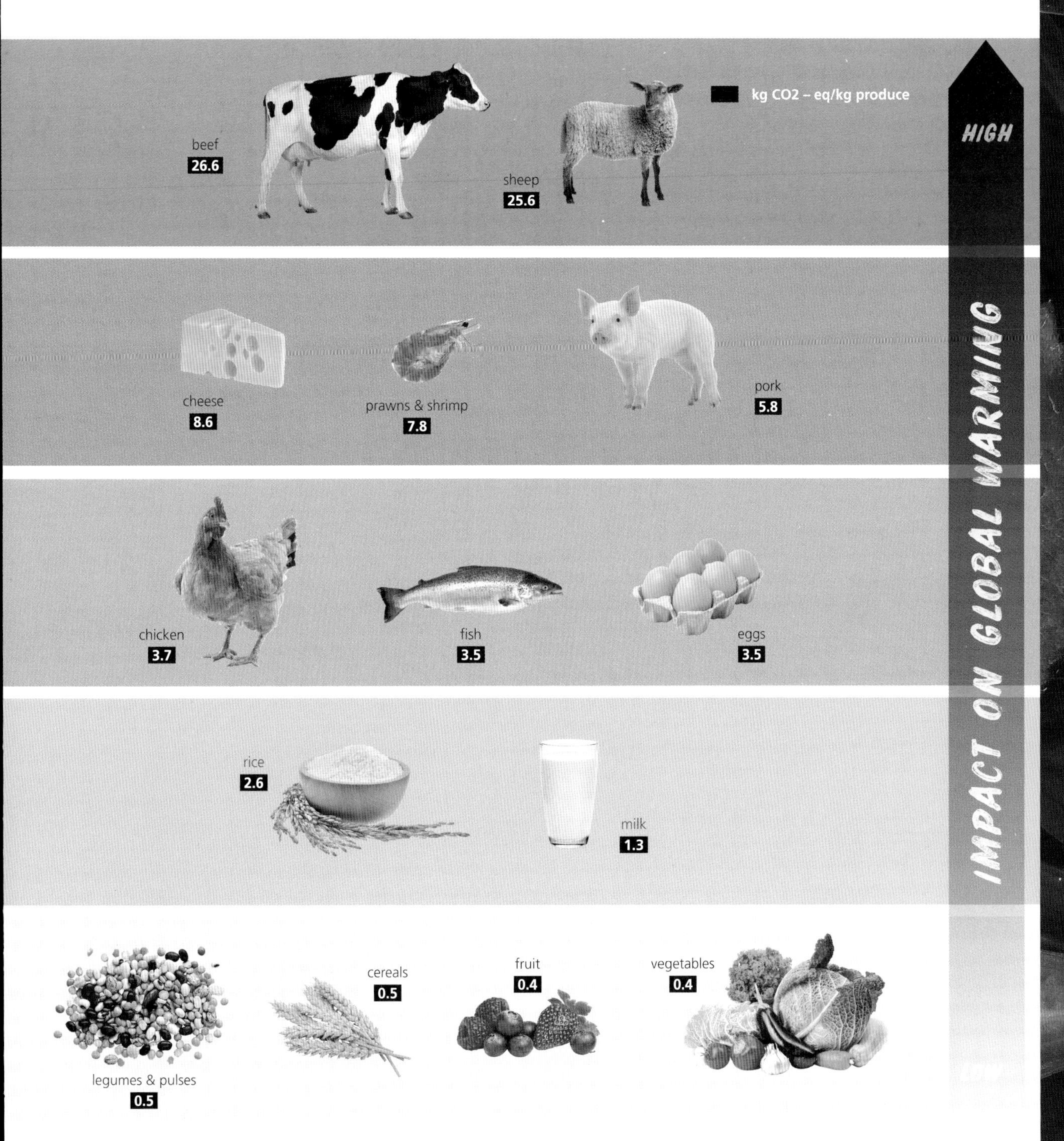

CHOOSE CARBON-SEQUESTERING FOODS

The reality here is that all plants sequester carbon, so the more plants we eat the better. But, as we have learnt, some plants can still damage our environment when they are grown in soil that is regularly tilled, when forests are cleared to plant them, and when they are treated with lots of chemicals. Easing off the annual crops and switching more to perennial staples would be a huge help. Perennial crops don't need to be pulled from the ground once or twice a year like most grains, sugar and soybeans do. This means they're terrific for our environment because they store carbon in soil, they require no tilling, often need fewer chemicals and use carbon to grow their yields. The bonus is they can also provide many of the proteins, fats and carbohydrates that we require.

Here is a list of plant foods that are *natural* carbon-sequesters. However, any gains are reduced or lost if the food is not sourced locally (since you're adding carbon from transport miles) and if you waste *any* of the food and it goes to landfill and produces methane.

Carbon-sequestering foods that are also good for the soil

HERBS + SPICES	VEGETABLES	FRUIT	OTHER
Chives	Asparagus	Apple	Almonds
Fennel	Broccoli (Nine Star or Purple Cape)	Apricot	Chestnuts
Garlic	Jerusalem artichoke	Avocado	Coconut
Ginger	Radicchio	Blackberries	Hazelnuts
Greek basil	Rhubarb	Cherries	Macadamias
Horseradish	Scarlet runner beans	Currants	Pecans
Lavender	Silverbeet	Dates	Pistachios
Lemon balm	Spring onions	Figs	Walnuts
Mint	Tree collards	Goji berries	Cacao
Oregano	Watercress	Grapes	Carob
Parsley	Yams	Kiwi fruit	
Rosemary		Lemon	
Sage		Lime	
Thyme		Mango	
		Nectarine	
		Olives	
		Orange	
		Peach	
		Pear	
		Persimmon	
		Plum	
		Raspberries	
		Strawberries	

GROW YOUR OWN FOOD

One of the best things you can do for the planet is to grow your own food. You can even turn your backyard into a carbon-sequestering food hub where every kilogram of vegetables you grow reduces emissions by roughly 2 kilograms (if you also compost organic waste). No pressure, but Eric Toensmeier's home garden is a tenth of an acre and it offsets the equivalent amount of emissions that an American adult uses in a year.

Carbon-sequestering veggie patch

These plants can all be grown in a temperate climate. For those in tropical climates, studies show that some home gardens can sequester more carbon than nearby forests. (You can plant bananas, mangoes, avocados and many other fruiting trees . . .)

WARRIGAL GREENS → cook and use like spinach or silverbeet; drought-tolerant (cook well to remove oxalic acid).

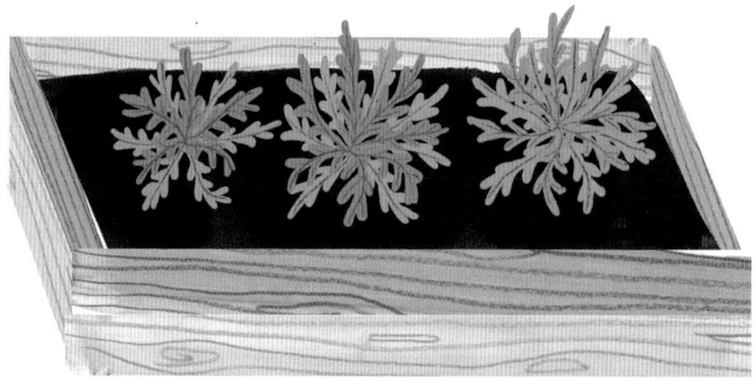

BANANA TREE → varieties include Lady Finger, Dwarf Ducasse and Pisang Ceylon.

WILD ROCKET → slower growing than regular rocket and drought-tolerant; listed as a weed in some areas so grow on edge of garden bed.

QUEENSLAND ARROWROOT → grows to 2 metres; root tastes like spuds.

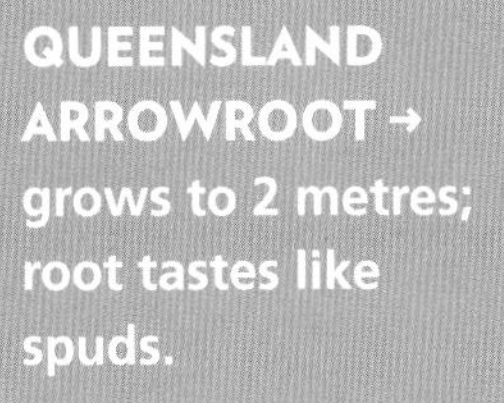

MINT → grow in a shady part of the plot; needs lots of water.

SCARLET RUNNER BEANS → best in cooler areas; spray blossoms with water in hot weather.

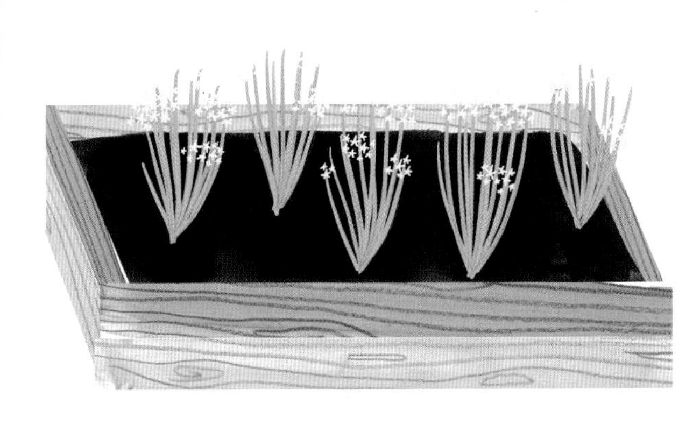

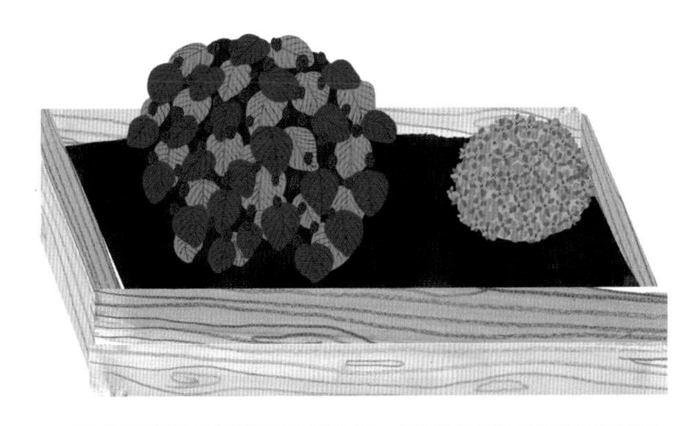

GARLIC CHIVES → drought-tolerant; harvest all year round.

RASPBERRIES or **BLUEBERRIES →** cooler climate.

GREEK BASIL → tiny leaves, fertilise well, harvest year round.

MONTAGUE
FRESH

'It's unbeatable to grow food in your own backyard. It's unbeatable to grow food all through and around your city. It's unbeatable to grow food on your roof. Those are the absolute gold standards of emissions.'

ERIC TOENSMEIER

Reduce food waste

Grow your own herbs

Or if you have to buy them fresh, chop and freeze leftovers in olive oil or wine.

Don't peel root veggies

Just give them a good wash.

Buy only what you need

Take a list and never shop when you're hungry!

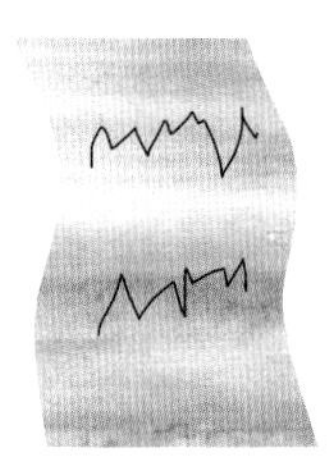

Cook ready-to-go meals in bulk

Store them in the fridge or freezer.

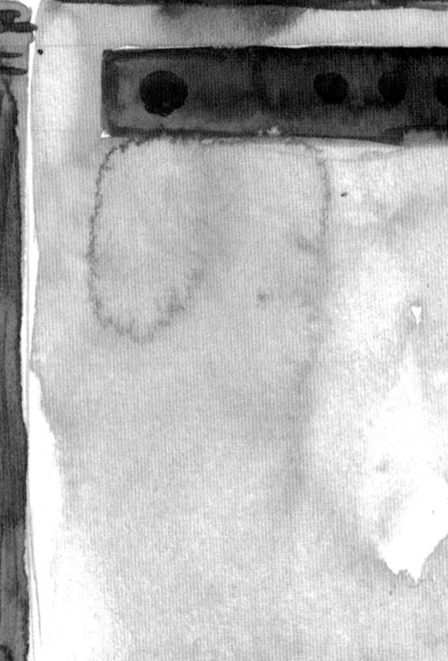

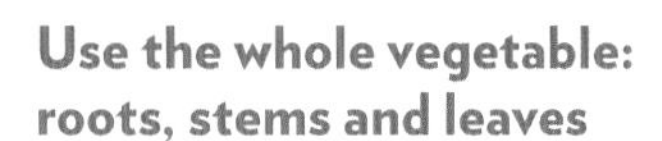

Use the whole vegetable: roots, stems and leaves

- Make 'rice' from broccoli or cauli stalks (or just cook them with the rest of the meal)
- Roast pumpkin skins
- Toss radish tops, beet greens and celery tops in salads.

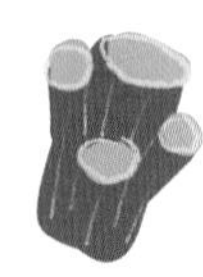

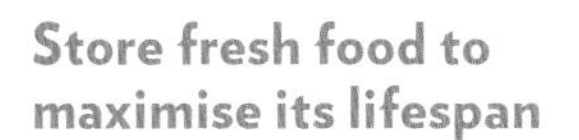

Store fresh food to maximise its lifespan

Keep spuds and onions in a cool, dark cupboard (plus other root veggies if your house is cool enough).

Store tomatoes, avocados and other fruit on the bench.

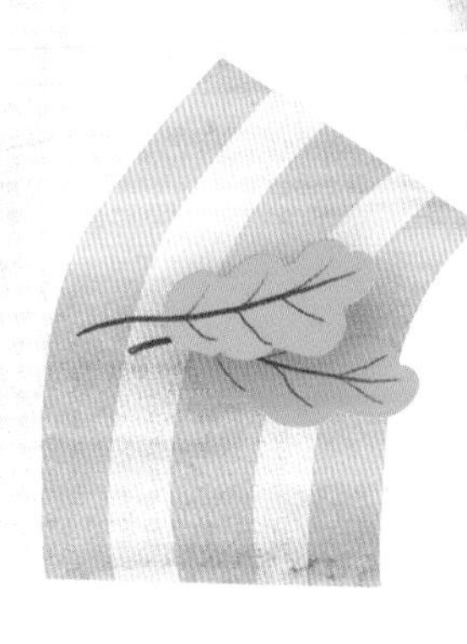

Loosely wrap leafy veggies in clean tea-towels and store in the crisper.

Store bread, nuts and seeds in the freezer.

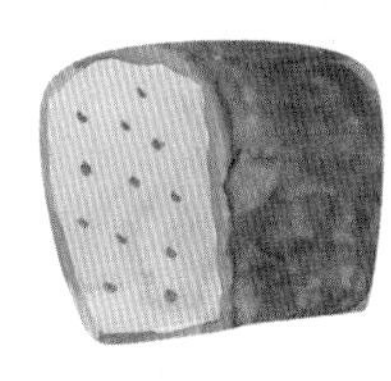

Encourage your council to collect veggie scraps

Our council now collects our food waste and turns it into compost. Can you organise a meeting in your area (perhaps after a screening of the film) and then collectively contact the council and suggest they do the same? They can make money from it by converting it to energy or selling it as compost. No-brainer.

START COMPOSTING AT HOME

If your council doesn't collect food waste, you can start composting at home. There are many ways to approach your composting but we have just embarked on a journey using **compostcentral.org**. It's a terrific resource. This practice involves placing a food bucket drilled with a few holes into our veggie patch. The bucket contains worms and other goodies. Once your food scraps go in, the worms take around three to five days to break it down, but they take it out of the holes and add it to your soil, greatly enhancing your veggie patch. The Compost Central website has a list of things that the clever worms can also break down including old socks and some bathroom products! Their gizzards are very good at turning all sorts of minerals and chemicals into healthy 'food-growing' soil.

Fun fact: earthworms are born with both male and female organs, and during mating both sets of organs are used by both worms. Talk about arousal. If all goes well, the eggs of both of the mates become fertilised and babies are born roughly every six weeks. This means you can increase the amount of food scraps in your compost every six weeks (more mouths to feed). We know some people who take home other people's food waste after dinner parties just to feed their worm army. 'Feed the worms not the landfill' could be a cool T-shirt by 2040. Or not. May still be incredibly daggy.

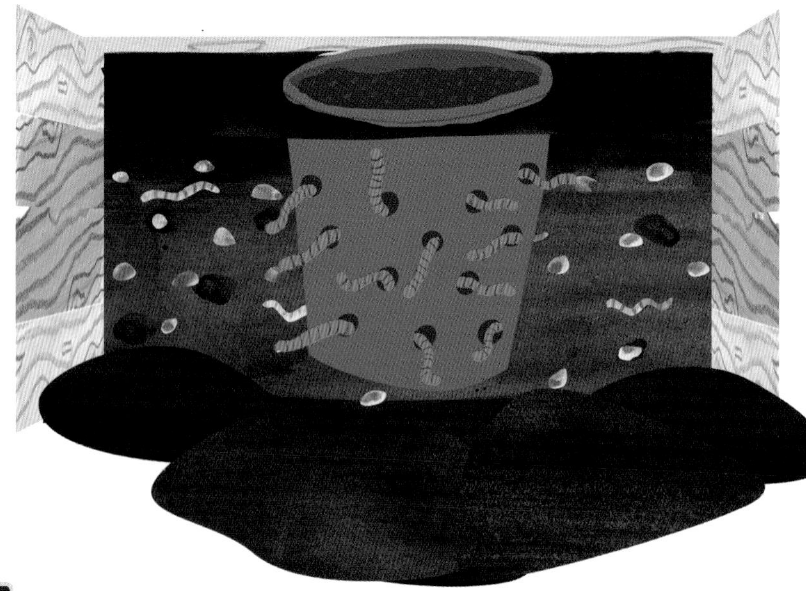

'The best way to use food waste, other than not generate it in the first place, is to actually get it back onto the land. Because that's where it came from. As a compost, it has really incredibly salutary effects on carbon sequestration, on productivity, on water retention.'
PAUL HAWKEN

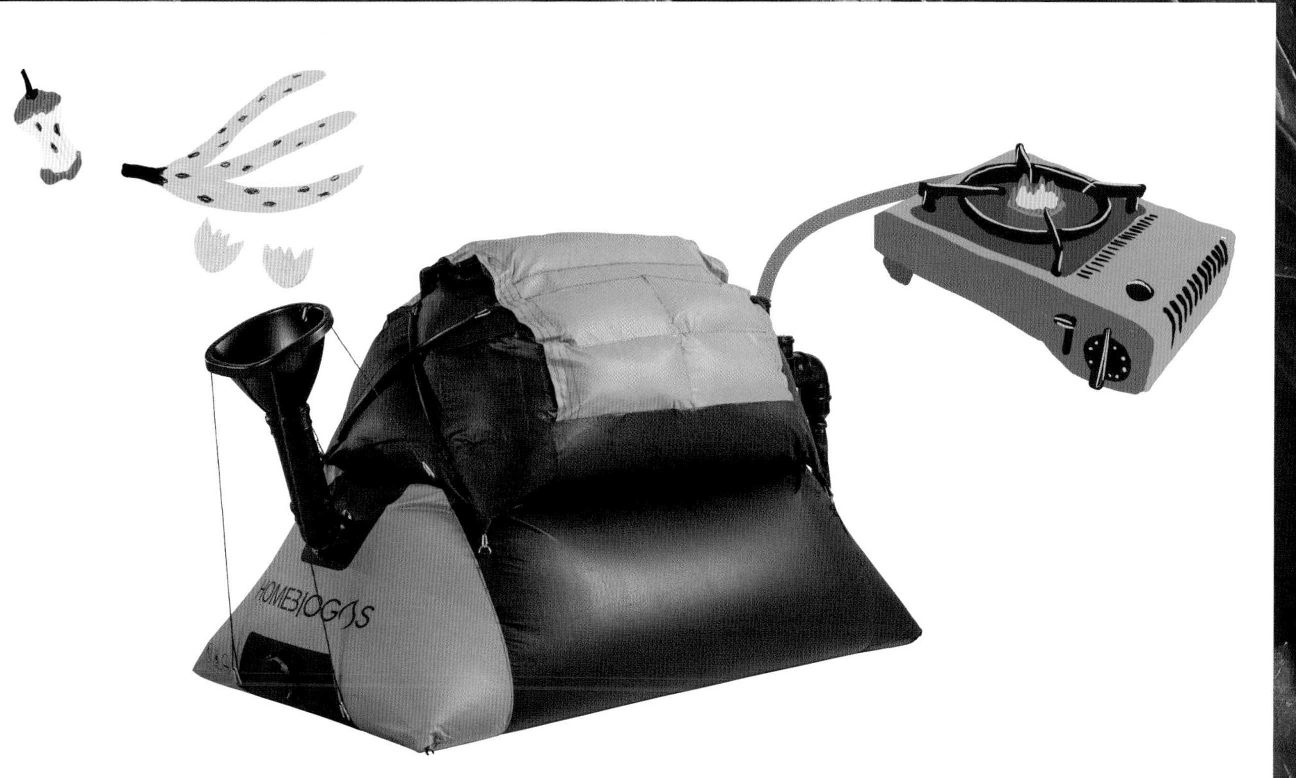

MAKE YOUR OWN BIOGAS

You can now purchase a home biogas kit from some Australian retailers or online (**homebiogas.com**). The kit allows you to convert your food and/or animal manure directly into a bio-methane gas that can be used to cook with (I have seen someone run hot water with it, too). It also provides a liquid fertiliser that will do wonders for your garden. It currently costs around A$800 and would be a great thing for a school or business to own, or for you to co-own with your neighbours (which I have recently done).

EMBRACE SEAWEED

Our daughter loves avocado rolled up in dried seaweed sheets as a snack (salty/fatty goodness). These two items combined make a great carbon sequesterer. They'll be even better when we source our seaweed from the local marine permaculture plot and grow our own avocadoes.

Professors in Shanghai have created kelp fibre spinning technology that could see kelp clothing enter the market in 2019. (At present, most seaweed-based eco fabrics use a small portion of seaweed fibre.) They say that 1 tonne of dried kelp could produce roughly 2000 square metres of seaweed 'fabric'. There are also many soaps, shampoos and other bathroom products now available that use seaweed as a primary ingredient: **seaweedbathco.com** is a great resource.

Check out **Evoware**'s edible seaweed plastic, too (evoware.id).

CONSU
PSA
PSA

MPTION
PSA
PSA
TT203 PSA

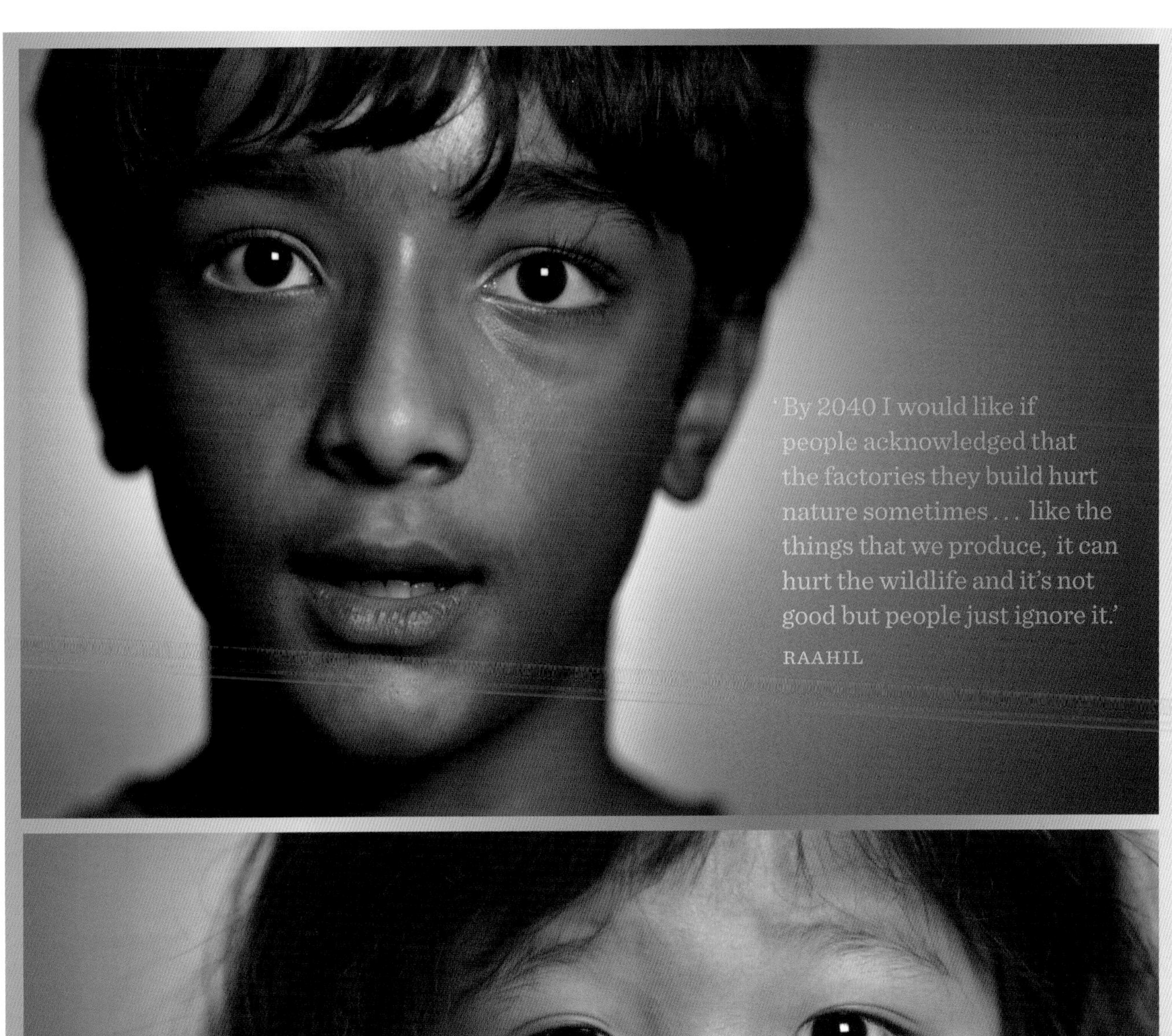

'By 2040 I would like if people acknowledged that the factories they build hurt nature sometimes . . . like the things that we produce, it can hurt the wildlife and it's not good but people just ignore it.'

RAAHIL

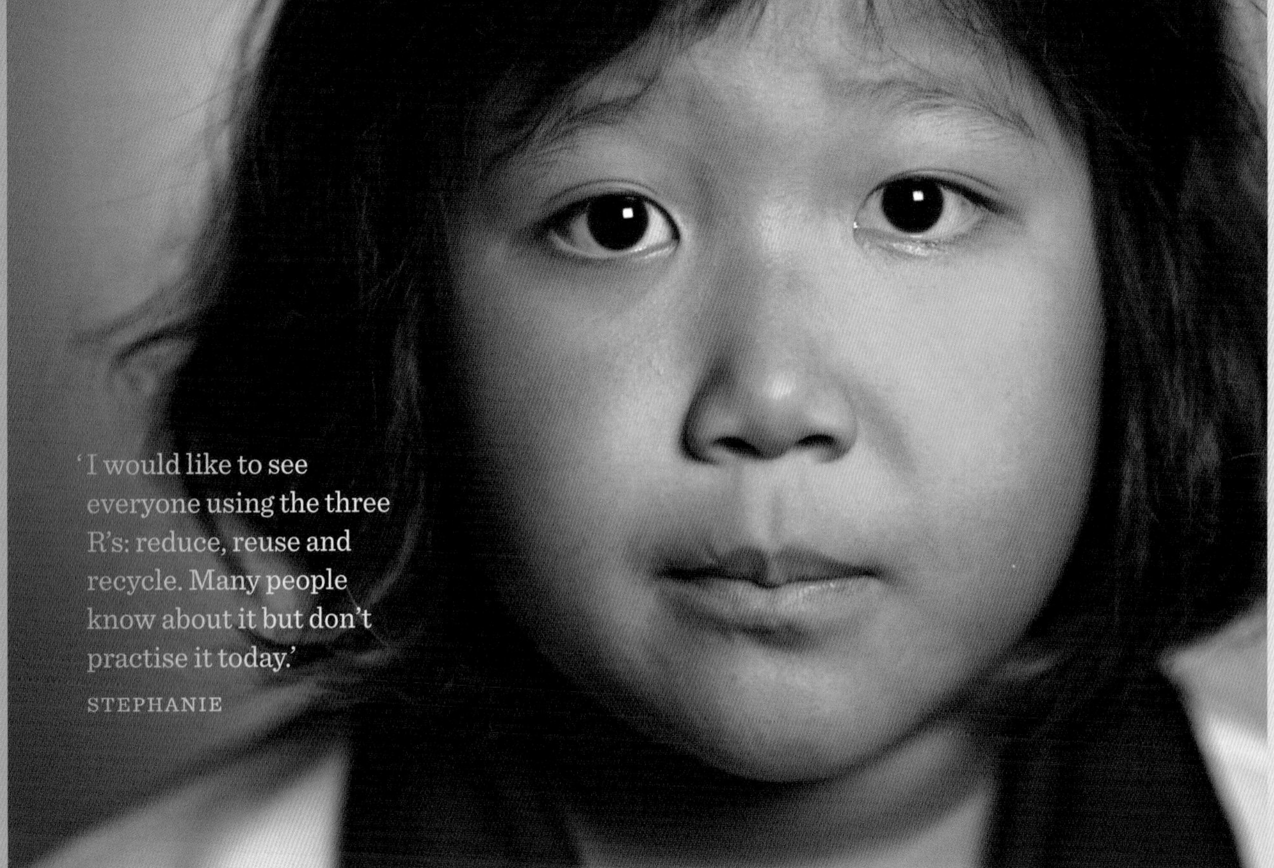

'I would like to see everyone using the three R's: reduce, reuse and recycle. Many people know about it but don't practise it today.'

STEPHANIE

It's at this point in the book that we need to have a slightly uncomfortable conversation. It would be really great if, by 2040, we could have global warming firmly on the path of reversal – cruising around in solar-powered, driverless shuttles, munching on regeneratively farmed produce while reading about the enormous seaweed plots that are pulling greenhouse gases from our atmosphere. Problem is, if we haven't curtailed our addiction to endless growth, we could also have more plastic in the ocean than fish and be running dangerously low on many of the resources that we take for granted today. There is a strong chance that we'd be seeing new tourist attractions such as 'Fast Fashion Mountain' – a spongy, climbable peak made up of discarded, worn-once clothing. Or perhaps you'd prefer a trip to the 'Apple Isle', a new island off Dubai made entirely of obsolete phones, iPads and laptops, its 'sand' comprising billions of white headphones with different-sized jacks.

As discussed earlier, a belief in endless growth was completely understandable when humans were emerging from the Great Depression or a World War. Economic growth has given many people across the globe wonderful and prolonged lives and should be acknowledged accordingly. The problem is, nothing in nature grows indefinitely; not a tree, not a mushroom, not a human tongue (that would be super-awkward). Things often grow very rapidly to begin with but then they mature and thrive. Our economists and politicians don't ever seem to address the endless growth conundrum. Is GDP supposed to just keep going up and up? Our economy constantly expanding? How is that even possible when we live on a planet with finite resources?

Economists often argue that endless growth is essential because it pulls people out of poverty. 'A rising tide lifts all boats,' they say. There's no doubt our growing economy has done miracles for some, but the miracles are occurring for fewer and fewer people. As I showed in the 'trickle down' diagram on page 60, just 5 per cent of the wealth created is making it to 60 per cent of humanity. So despite the last few beneficial decades, we still have 4.2 billion people living on less than $5 a day.

David Woodward at the World Economics Association did some sobering maths around this and found that, at our current growth rates, it would take 207 years to pull those 4.2 billion people above $5 a day, and the economy would have to grow 175 times its present size. This would not only create unimaginable levels of income inequality but would also wipe out every living thing on the planet.

Questioning the endless growth narrative is our toughest collective conversation because it is now so intertwined with our culture. Political parties are shaped by it and even the threat of it slowing or stopping can lead to banks failing, businesses going bankrupt and people losing their jobs. The economist Kate Raworth explained it to me well: 'We have to understand why and how we've become addicted to growth. Over the past couple of centuries, our economies have been structured so that they demand, depend upon and expect unending growth in national income – financially because at the heart of the financial system is the constant demand for the maximum rate of return, and politically because governments want higher tax revenues but no one wants to raise taxes. Well, the way to do that is to grow. Plus no one wants to lose their place in the G20 family photo because if any one economy stopped growing, they'd be booted out by the next emerging powerhouse. So they've all got to grow, to keep up with everybody else who's growing. Then socially, we're addicted to growth because after a century of consumerist propaganda, we've been convinced that the best form of therapy is retail therapy. We need to overcome these addictions, but I don't think any of them are insurmountable.'

It is interesting that after the events of 9/11 in New York, the Bush and Blair administrations in the US and UK didn't encourage people to spend time with their loved ones and connect, they urged people to go out and shop. Spending money and increasing GDP was the best way to fight back. This is a slightly different approach to the climate scientists who completed the IPCC report I mentioned earlier. They believe we have to scale down global material consumption by at least 20 per cent (with rich countries leading the way) if we are to avoid some serious climate pain.

This doesn't necessarily mean we should be anti-growth. We want lots of growth in the things that are going to cure our sick planet, but the expectation that they will continue to grow beyond that point defies the laws of nature. Because we are overshooting the resource allowance that the earth affords us each year (we need 1.7 Earths' worth of resources for a single year) we are in effect, stealing money from our children's money boxes.

The IPCC report asks us to lean more towards a 'circular' economy where more of our materials and resources are shared and reused. Car-sharing is one of those ways. And there are many other terrific initiatives, such as community tool sheds that allow various power tools to be rented or used

free of charge from a central hub, rather than everybody in the neighbourhood owning an individual tool that lays unused for the majority of the time. There is also the Cloud of Goods organisation that allows scooters, strollers and other items to be rented and delivered to customers' homes.

Our rapacious appetite for growth is not only impacting our resources, but also our social stability. Everything is viewed as a commodity to be bought or sold – childcare, education, health and even prisons in some places. Something is seriously wrong when prisons are trying to admit more people than they are releasing for reasons of profit and economic growth. The growth machine must eat and it doesn't care what it puts in its mouth.

It would be easy at this point to hurl all the blame at politicians, or perhaps at trade treaties, financial institutions or even a shadowy group of illuminati/Bilderberg types if that's your thing. But I also think we need to take some responsibility ourselves. Why do we think we need more and more? Is this just part of our nature? Or are we genuine victims of a clever and brilliantly orchestrated advertising paradigm?

In seeking to understand what drives our consumption, and therefore how to manage it, I have researched multiple arguments from all sides. I'll look at some of those later, but first share my own insights.

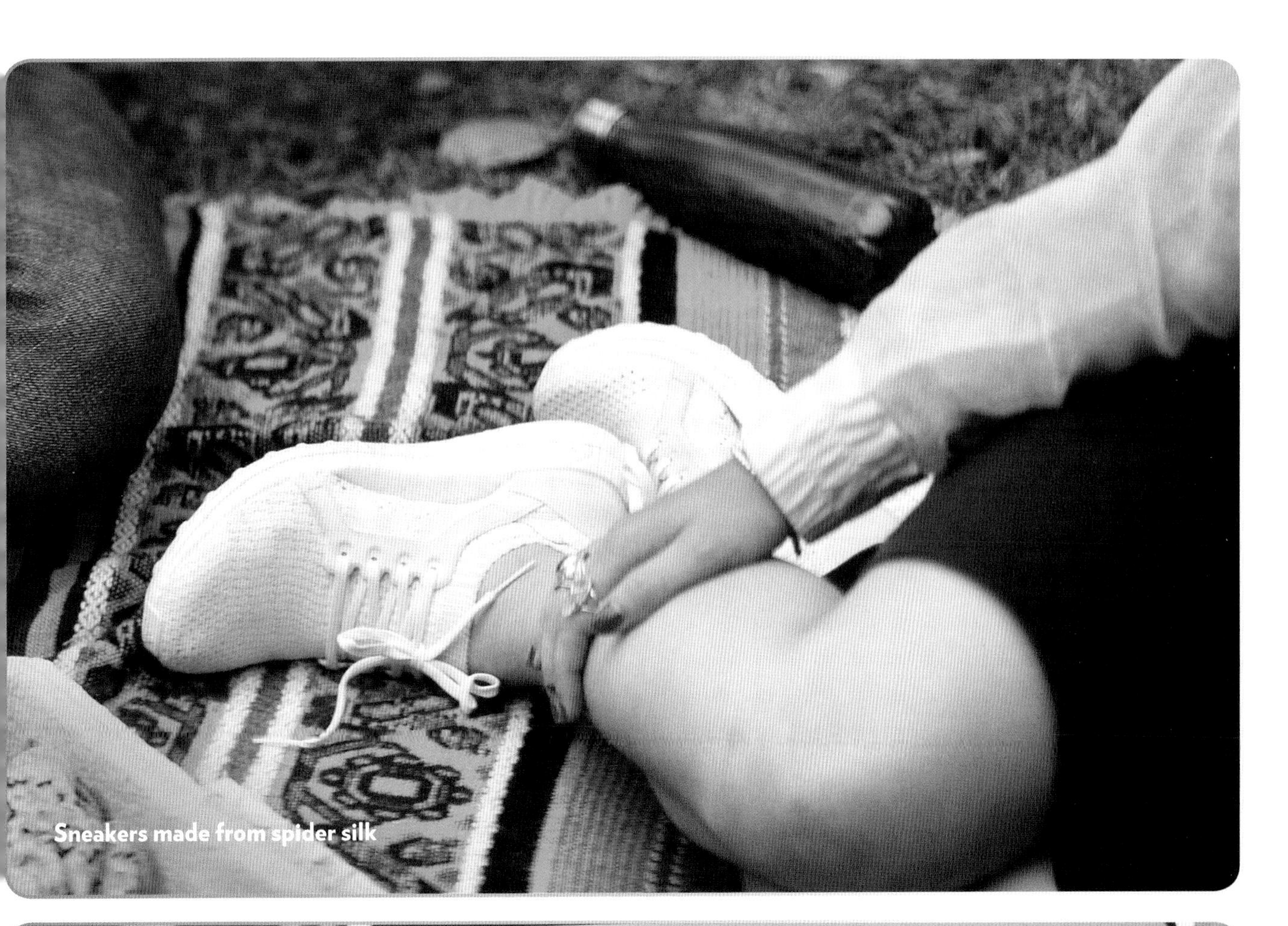

Sneakers made from spider silk

LOCAL SEED CUP?
PLEASE LITTER HERE
WE ARE REFORESTING

Genuinely biodegradable coffee cups
that contain a seed (specific to the region);
the cup is buried – and a tree grows.

THE CRAVING FOR *CONNECTION*

Our daughter is four years old and has already taught me more about myself than a couple of trips to India, an Amazonian ayahuasca experience, any Eckhart Tolle book or a ten-day silent retreat. Oh yes, I have drunk deeply from the well of self-exploration (and still occasionally feel thirsty).

You see, when our daughter gets upset, we know that her needs aren't being met, but it's really only in the last year that we have understood that 99 per cent of the time those needs involve connection. Yes, it can be tiredness and hunger too, but she is her happiest when one of us is present with her, our phones are down and we are fully engaged. She feels seen, heard and understood. The trouble arises when she feels she hasn't had enough of that; particularly when Dad has been on his phone or the computer for too long. Now obviously we cannot be truly present with her all the time, but we notice that when we regularly interact with her in that way she is more settled. It's as if her cup gets filled, and this allows her to get on with doing her own thing.

I do wonder how many of us still have a toddler inside of us with unmet needs? As we reduce our face-to-face encounters, what impact is this having on our sense of connection? Could we be turning to more sugary foods or drinks

because they release the same opioids that love does? Could we increasingly find ourselves online seeking 'likes' from our posts or joining a 'tribe' of like-minded people because we are lacking genuine connection in our daily lives? Does the rise of reality TV and celebrity obsession reflect our belief that our lives are not enjoyable enough?

I ask these questions from direct experience. When I am working away from my family, it's amazing how often I find myself slipping into a pattern of eating foods that aren't that great for me or being glued to my phone trawling the internet. Am I trying to 'fill my cup' and meet my 'unmet needs' in different ways? Interestingly, my wife says she feels the urge to 'shop' more when I am out of town.

Perhaps a major part of healing our ecological ailments actually comes down to deepening our connections with each other. Balancing our time online with more face-to-face meetings, joining social groups, caring for animals, caring for the soil. These could be the very things that help fill a gap that might otherwise be filled with a need to consume more and more and thus diminish our resources.

I recently read a book called *Rules for Revolutionaries*, which documents the Bernie Sanders campaign in the lead-up to the 2016 US election. What was different about this campaign was that volunteers didn't want to do the usual cold calling or emailing, they wanted to organise face-to-face meetings. People would attend a gathering, feel heard and part of a community, then help organise a different gathering, filling a room with people connected by the same shared experience. The cycle would continue and the campaign quickly became a force. A huge part of why I made *2040* was so it could be screened and discussed in large groups at schools, businesses, universities, community town halls and government functions. The only way we are going to reach a better future is by coming together, preferably in person, to hatch a plan.

'People regain a vitality and a joy through the connection to nature and to community. That's how we evolved. We evolved close to one another, and close to the animals, to the plants, to the life all around us. That's who we are.'

HELENA NORBERG-HODGE

But, apart from just connecting with each other on a genuine level a hell of a lot more, there are some pragmatic, less Byron Bay-inspired solutions for reducing our material consumption. They are a blend of business innovation and government intervention.

The economist Kate Raworth has an idea of using taxes as a lever to add value to our resources. 'At the moment, businesses are incentivised to hire as few people as possible, because they pay payroll taxes. They are also incentivised to use as many resources as possible because there's almost no taxes against that. We need to flip that around. So instead of penalising businesses for hiring more people, penalise them for using new resources and sending stuff to landfill.'

I can just imagine this tax leading to organisations that use a lot of plastic sending out boats to collect ocean plastics and reuse them or having much more efficient recycling programs. What a brilliant way to rapidly clean up our mess (and create new jobs).

Another economist, Richard Denniss, believes we should become more materialistic, in that we should value materials more. A true materialist wouldn't wear a garment just a few times and then throw it away. They would savour and appreciate it and get it repaired. More and more companies are beginning to offer this service. Patagonia and Nudie jeans now allow you to bring in your item and have it repaired instead of simply replacing it. There are also online services that allow you to 'rent' clothes for a certain period of time. You return it when you are done, they clean and/or repair the item, and off it goes to someone else who wants it. I have recently found myself becoming increasingly 'materialistic' (but haven't yet found the confidence to say it out loud).

I found many wonderful examples of individuals and organisations embracing the 'circular economy' approach . . .

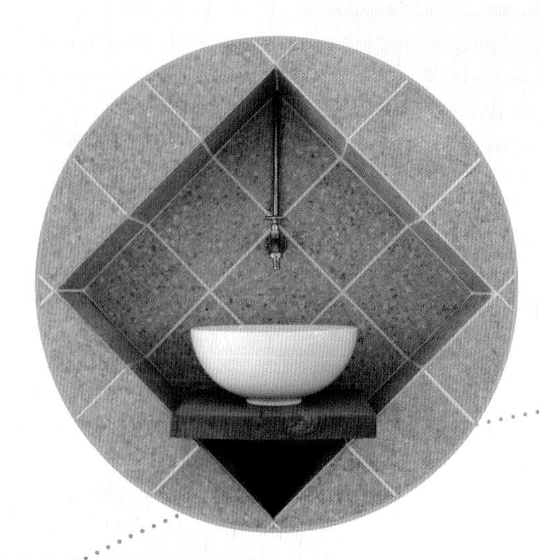

Wall tiles made from plastic bottles.

Containers or lamps 'grown' from mushrooms, which are often stronger than concrete.

Plastic containers made from food waste or seaweed.

A data centre in Sweden runs pipes past their servers to heat water and then sends it back to the suburbs.

A girl in the Netherlands makes clutch bags from fermented tea and colours them and her jewellery with vibrant 'bio inks' found naturally in certain soil bacteria.

One company in Rotterdam fits out a soccer stadium with tiles made from recycled beer cups collected after games.

There are ideas like these springing up across the globe (I list a few more in 'More reasons for hope'). We just need to give them greater attention and support so that they can become the norm by 2040.

'I want people to stop messing with the Earth.'

PHAROAH

'Well, a lot of people text. Maybe we should talk face to face more. I'd probably like to see people less on electronics. Maybe you could invent a different electronic that would help people learn but also be fun.'

CHARLOTTE

ENVIRONMENTAL *DASHBOARDS*

For the majority of our evolution, we have been deeply connected to our resources. We carried water up from the local stream to bathe or drink, we chopped our own wood to keep us warm or to use for cooking, and we handmade our clothes (which usually involved shearing a sheep or skinning an animal after we ate it). Today we rarely pause to consider our resource consumption. We turn on a tap or flick a switch and take for granted that the resource will be there.

I found myself in the town of Oberlin in Ohio. A town famous not only for its integrated communities that allowed blacks to attend its college in 1835 but also for the 'Anti-Saloon League', which became one of the most prominent lobby groups for prohibition (the banning of alcohol between 1920 and 1933).

After a refreshing and slightly mischievous glass of beer, the film's director of photography (Hugh Miller, superb job) and myself went to meet Dr John E. Petersen, chair of environmental studies at Oberlin College. John has teamed up with the local council to place 'environmental dashboards' throughout the community.

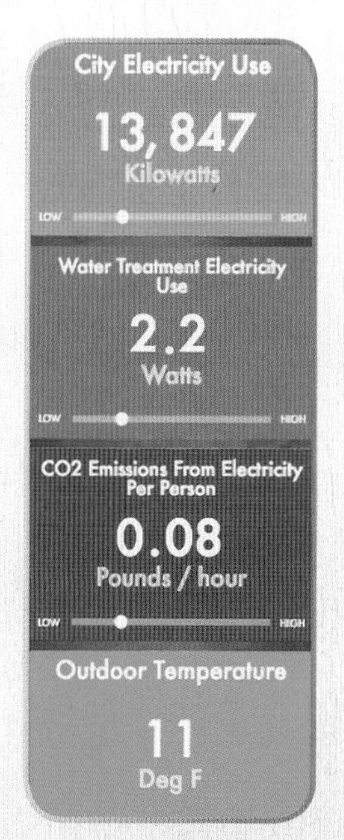

As Hugh and I walked down the main street, we noticed a dashboard in a cafe window, another at the local library and in the lobby of the hotel where we were meeting John. The environmental dashboards display the energy and water use of the buildings in which they are placed. In addition to these downtown buildings, there are also dashboards in 26 dormitories on the Oberlin College campus, and in four local schools.

We went to see the dashboards in action at one of the schools. Talking to some of the kids, it was obvious that they had developed a passion and, quite surprisingly, an emotional response to their school's energy use (I have had a similar response to my own energy bills at times). This has been aided by a cartoon character on the dashboard in their classroom that is either happy or sad depending on consumption. 'We have done controlled studies with a neighbouring school that has the same curriculum but doesn't use the dashboards,' John told me. 'We have found that our students experience more emotions around resource use and, when you ask them about resources, they use the word "we" instead of the word "I", so we are seeing very real changes in their behaviour.'

The kids have now set up an Eco-Olympics where they compete with the other schools to win the 'highest reduction in energy and water use'. Very 2040.

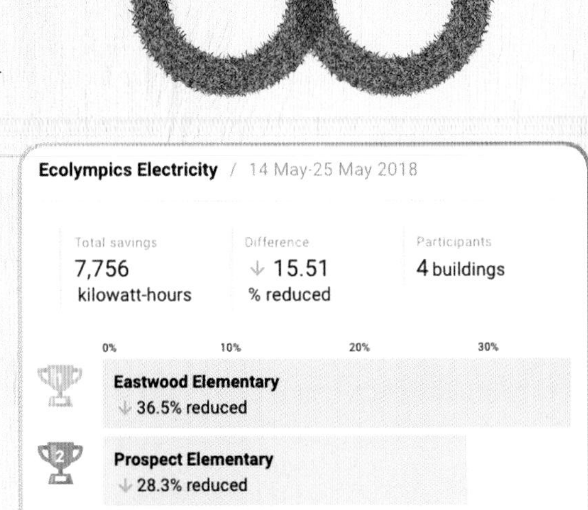

Imagine how this technology could be applied to things like food waste levels in landfill, or rubbish and plastic levels in our waterways. The data could be sent directly to your phone by the local council, displayed on prominent websites or shared as daily news. This could shape our behaviour around a range of resources. When people are given the correct information they can make an informed decision. As John discovered with the Oberlin schoolkids, this information can also help shift our mentality from one based around individual consumption, to one that sees how that consumption affects the whole.

In the film, I propose that by 2040 many advertising billboards (which are currently used to promote growth and consumption) could be replaced with various environmental dashboards. One in Times Square displays New York residents' daily meat consumption. People may be more conscious of ordering a steak or the size of that steak if they know exactly where the city's level of sustainable meat consumption is for that day.

This resource education will become more and more important as the world fully comprehends our overshoot dilemma. But there's another aspect of education that could have an even more profound impact . . .

Electricity
Water
Streams
Weather
1.8
0.07

NEW DAY
TUESDAYS
9/26 FOX5
MARRIOTT
MARQUIS
NY MEAT CONSUMPTION
PER PERSON
85G
A DAY
NY
IS PROUD TO BE A
REGEN
CITY
target under 65g a day

FOOD
Sweden
UK
Norway

LEICESTER SQU

If I had to name the moment of this adventure where I felt the most gobsmacked, it was when Paul Hawken told me that the number one solution to reversing global warming is the empowerment of women and girls (by combining solutions number 6 and 7; see page 32.)

As I discuss in the film, I take for granted that our daughter will get to complete her education. Unless something unforeseen occurs, she will be able to finish her schooling surrounded by new friends and an expanded mind that is ready for the opportunities that await. Yet millions of girls around the world don't get that opportunity. 'Educating girls is so interesting,' Paul told me. 'Basically, about 98 million girls are kept out of school after a certain level and put to work to earn money to put their brothers through school. Or for early marriage. For whatever reason, that girl becomes a woman on somebody else's terms. Culture, village, family, religion – who knows? A combination. Her average reproduction rate is five-plus children. If she is allowed to go to school and matriculate to what we call high school, she has an average of two-plus children.'

I spoke to Dr Amanda Cahill, CEO of the Next Economy, about this. She explained that high reproductive rates in the absence of education is not confined to poorer nations but also occurs in wealthier nations. She also stressed that this wasn't simply about providing more education; girls also need access to health and reproductive services plus decent work opportunities. It is when these things align that a girl will choose to have a child when she is ready, and will usually have fewer children.

There are obviously huge advantages to educating girls. Not only are more women empowered to enjoy a better quality of life, but the child of that girl is much more likely to receive an education, thus breaking the 'cycle of poverty'. When it comes to the impact on the planet, the population decreases, which puts less pressure on resources like land, water, energy and other materials.

Paul told me that completing the education of those 98 million girls would make a significant difference to population growth. According to the UN's population predictions, it would mean 1.1 billion fewer people on the planet. 'It's a form of family planning,' he explained, 'but it's not coercive. It's not control. It's just empowering girls to be who they want to be, and you have these incredible benefits.'

MORE REASONS FOR
HOPE

According to a recent Yale study, 70 per cent of people in the US now believe protecting the environment is more important than economic growth.

In Bangladesh, the fertility rate (average births per woman) has dropped from six in the 1980s to two due to education programs, social support and shifting norms (including by changing what an 'average family' looks like on popular television and radio shows).

Scientists have identified that just 10 rivers in the world are responsible for over 90 per cent of the ocean's plastics, and 8 of these rivers are in Asia. Research is now focused on how to collect the plastic before it reaches the ocean.

Many governments are beginning to understand that the plastic must be stopped at the source. Costa Rica has announced it will ban all single-use plastics (straws, bottles etc) by 2021. India will do the same by 2022.

A beachside cafe in the US offers free coffee to anyone who collects a bucketful of plastic from the beach. Your turn, Australia.

A company I visited in the Netherlands has begun using recycled plastic to make durable roads and bike paths. They are stronger than bitumen roads and much longer lasting (and have been tested to make sure no micro-plastics can break away and wash into waterways).

Adidas has agreed to only use recycled plastics in its products by 2024.

A city in Colorado is processing 8 million gallons of human waste per year into renewable natural gas (also known as bio-methane). This is being used to fuel fleet vehicles, including garbage trucks, street sweepers, dump trucks and transit buses (and no, it doesn't smell, 'cos I know you thought that).

BOSS Menswear now offers high-quality vegan shoes made of pineapple leather. The material, known as Piñatex, is 'an innovative leather-alternative created from pineapple leaf fibres'. Because the leaves are a by-product of harvesting the fruit, no additional resources are needed to produce the material. This means that farmers can receive an additional source of income.

WHAT YOU CAN DO TO HELP

There are so many people doing incredible things in this space right now. From #zerowasters on Instagram to websites that facilitate clothes sharing and swapping, the circular economy movement is well underway.

Be a true 'materialist' – value your things

A couple of years ago, my wife, Zoe, read *The Life-Changing Magic of Tidying Up* by Marie Kondo. It is all about simplifying and decluttering your life, plus truly valuing and respecting the things you do own. Arguments my wife and I used to have about certain purchases have disappeared. She is a conscious, thoughtful shopper and really loves the things she has. Plus our shared wardrobe now has so much more room!

This 'materialism' can extend to your morning coffee. Take a cup to the cafe (any kind – I often take my own mug). I even take my own cup on flights (so much waste at 30,000 feet!). It's really not that hard, no matter how busy you are. Fortunately, this is an area where cultural norms are shifting fast. It won't be too long before it's seen as embarrassing to be caught holding a takeaway coffee cup or plastic bottle of water (it already is in some parts of the country).

Dress for the planet

Find out where your clothes come from, how they are made, and the conditions of the workers (I had my eyes prised wide open when visiting a well-known western brand's factory in Bangladesh).

You might be interested in supporting companies like **ShareWear** that allow people to borrow clothing items for free if they agree to share the items forward after using them for one week. This is done via social media and sharing sites. This is part of 'closing the loop' in the fashion industry.

If you wish to explore more in the fashion space these websites are great:

- **sharewear.se**
- **fashionforgood.com**
- **circularfashion.com**
- **reloopingfashion.org**

Choose environmentally friendly fabrics

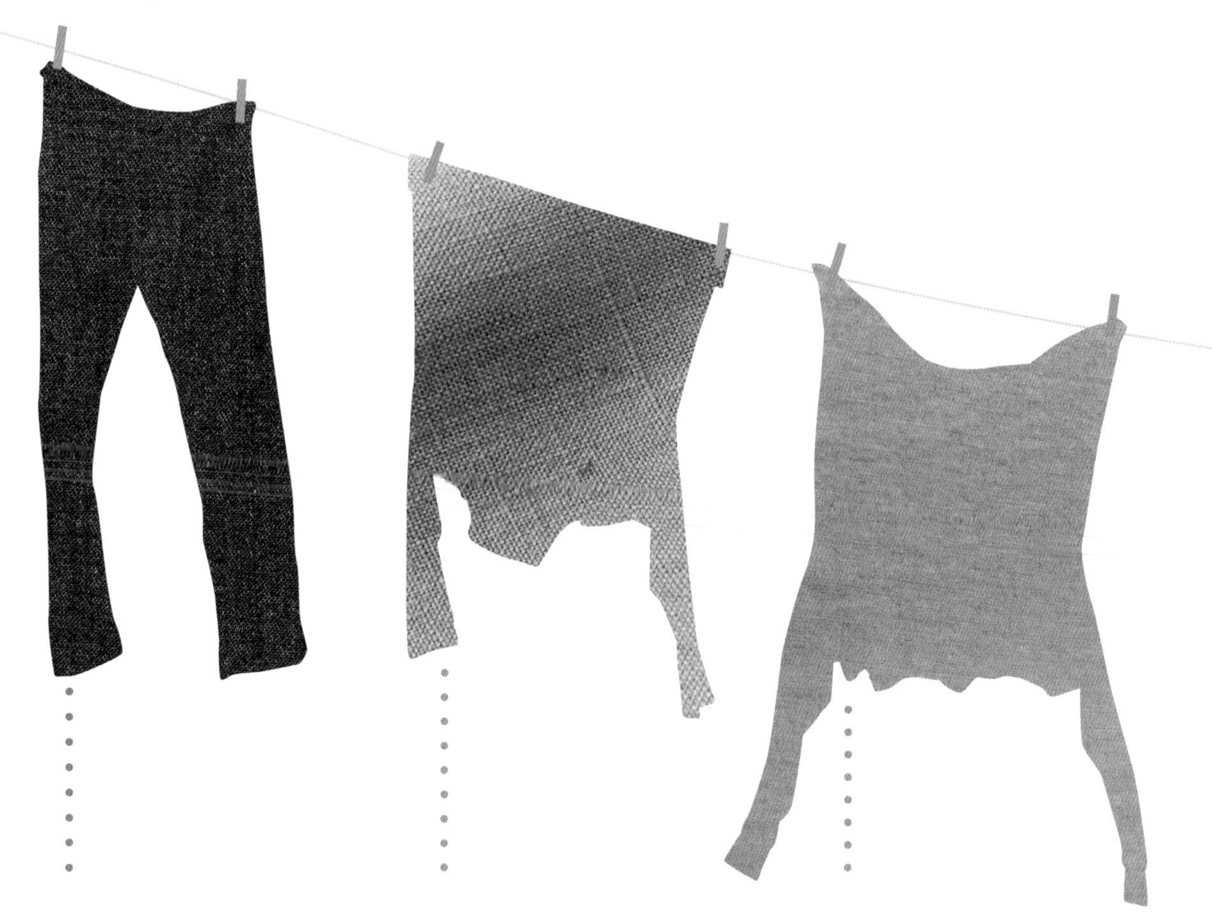

GRADE A+

Tencel (lyocell) is made from wood pulp sourced from fast growing, carbon sequestering, sustainable eucalyptus tree farms. It is biodegradable and does not need to be bleached before dyeing, due to its pure white colour at production. It's soft, breathable, lightweight, and comfortable.

GRADE A

Hemp fabric is made from a species of *Cannabis sativa*, a hardy, fast-growing, naturally pest-resistant plant with high yields per hectare. Like linen, however, it is an annual and requires tilling of the soil. Let's hope this can be grown with regenerative farming methods in the future.

GRADE A

Linen is created from the fibres of the flax plant. Flax is a plant that grows worldwide and the production process is quite simple and sustainable, which is one reason why linen has been used for so long. It's light and comfortable and has been used as an ayurvedic healing fabric in India for centuries. Organic linen is preferable due to the agricultural practices that are applied.

GRADE B+

Silk is an animal protein fibre produced from the cocoons of silkworms, bees, beetles, silverfish and mayflies. In conventional silk production, a silkworm's lifespan is cut short by immersing the cocoons in boiling water, thereby killing the pupa inside to keep the thread intact. (Suddenly silk lingerie isn't quite so sexy. Is that boiled pupa you're wearing?) While it does provide income to those who grow the silkworms, some people have issues with the killing of the moth in production. What is gaining in popularity is 'peace silk', or Ahimsa silk, which is made without killing the creatures that create it.

NOTE: *Compared to synthetic materials like nylon or polyester, silk is a biodegradable material so is more sustainable and thus gets a higher grade.*

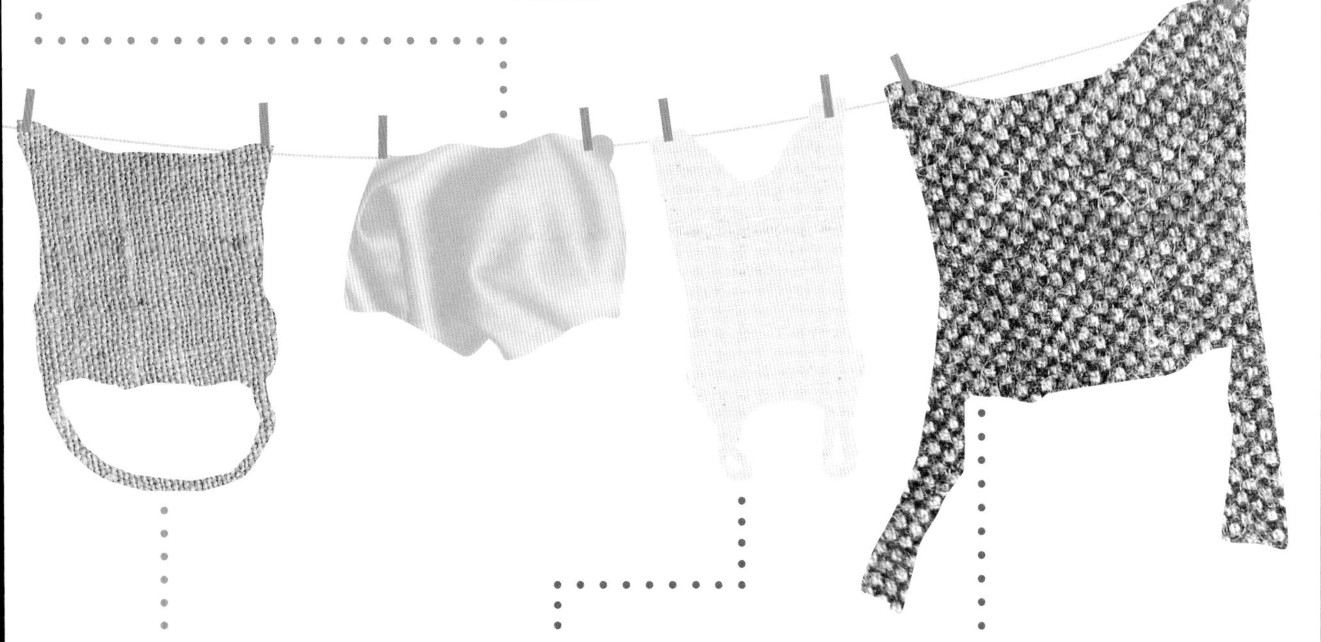

GRADE A

Jute (hessian) is made from plants in the mallow family. It's very strong and rough, so not used for clothing – more for bags, twine, rugs, pet beds etc.

GRADE B

Organic cotton is grown without the use of pesticides or fertilisers. Farmers use crop rotation, composting, and natural pesticides such as castor oil sticky traps. This also reduces the need for water. (Non-organic cotton is a thirsty crop requiring more than 20,000 litres of water to produce 1 kg of cotton.)

GRADE B

Organic wool is from sheep that have been fed only certified organic feed and forage; are free of synthetic pesticides, hormones, vaccinations, and genetic engineering; are kept in humane conditions and shorn using practices that encourage livestock health.

NOTE: *Fabrics from animal farming operations have a higher carbon footprint than plant-based fabrics.*

Shower and flush for the planet

Household water for showers, watering the lawns, flushing the toilet or doing the laundry all use substantial energy because water is very heavy and it needs to be pumped for treatment, storage and distribution. Creating hot water in our homes is responsible for a quarter of residential energy use worldwide! A big help is using low-flush toilets, water-efficient washing machines or dishwashers, and more efficient shower heads (and having shorter showers). These combined can lower household water use by 45 per cent. According to the US EPA, if just one household in 100 swapped an old toilet for a more efficient new one, it would save enough energy to power 43,000 US households for a month.

Internet search for the planet

When you search using Google, the company makes money by auctioning off your interests so you can get bombarded with tailored advertising. The search engine **Ecosia** is downloadable in seconds (I use it for all my searches) and instead of keeping that ad money for itself, it uses it to plant trees. Fifty million trees have been planted as I write this and the company has just bought a forest in Germany to protect it from logging.

Help educate girls

There are many not-for-profit organisations working to educate girls that you could reach out to (see **educatinggirlsmatters.org**). Perhaps the best known of these organisations is **malala.org**. Founded by the international human rights icon Malala Yousafzai, the fund works directly with girls in Pakistan, Afghanistan, India and Nigeria, as well as countries housing Syrian refugees, such as Lebanon and Jordan. Donations to the fund are used to invest in schools and supplies, as well as place activists and educators in the girls' communities. Other organisations include:

- **Camfed (Campaign for Female Education)**
- **Girls Education International**
- **Global Campaign for Education**
- **Global Fund for Women**
- **Mama Cash**
- **Girl Effect**
- **Girls' Empowerment Initiative**

Unfortunately, providing teachers and classrooms is only part of the solution, as there can be social and cultural issues that interfere with education. For example, every year 15 million girls under the age of 18 are married (see **girlsnotbrides.org**), or there can be taboos and traditions surrounding menstruation that mean girls are afraid to go to school (see **period.org**). Even worse, gender-based violence (from other students and even teachers) can make it unsafe for girls to attend school (**girleffect.org** is working to change this). As you can see, this is an area that could do with a lot more support.

Malala Yousafzai

Shop at bulk-food stores

These stores are great for buying staples such as flour, dried herbs and spices, nuts, seeds and dried legumes as they don't use any packaging (single-use food or drink packaging is a primary source of ocean plastic). Just take your own repurposed jars and containers, or use the reusable paper bags they provide. See **thesourcebulkfoods.com.au**.

Your personal Earth Overshoot Day is:
06 Sep
If everyone lived like you, we would need
1.5 Earths

Calculate your global footprint

The Footprint Calculator (**footprintcalculator.org**) is a terrific, clear tool that shows you how many Earths' worth of resources you are currently using in your own life. I was stunned by the difference less air travel would make to my own footprint. It's a fun exercise to do with children as the graphics are very clear and playful.

Hold a 'What's Your 2040?' night

Head to our website – **whatsyour2040.com** – to see how you can not only screen the film at your home, school, workplace, university or community hall, but also host a conversation afterwards about what your 2040 could look like. We have designed a simple and clear process so that anyone (no matter how much experience you've had with public speaking) can run the event and get important debates and discussions underway. This is a great way to actively 'join the regeneration'.

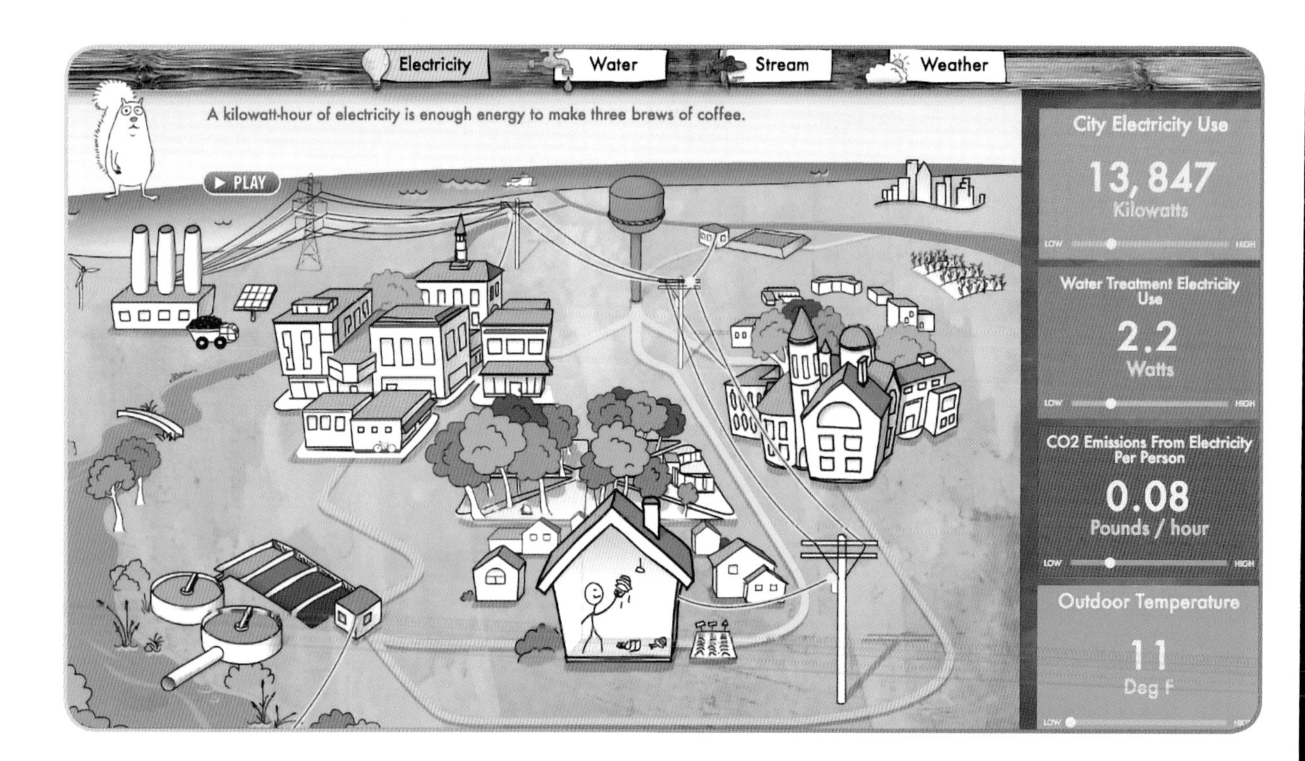

Create an environmental dashboard at your school or workplace

After you show *2040* at your local school, workplace or in your community, you might want to pitch the idea of creating an environmental dashboard to display resource consumption. Here's a website that explains how you can make that happen: **environmentaldashboard.org/ bring-dashboard-to-your-community**.

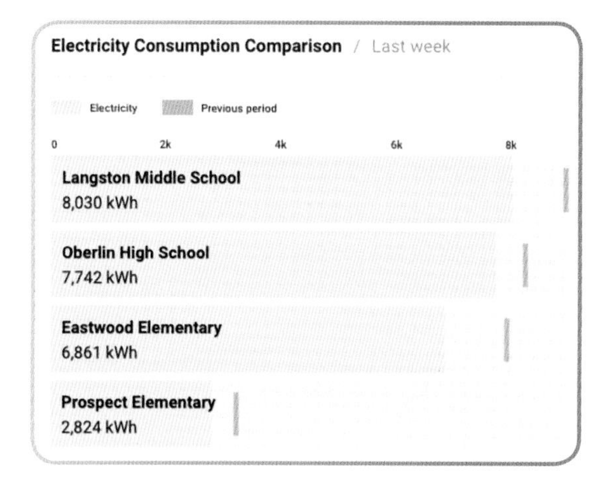

Wise up on biodegradability

Unfortunately, some plastics companies are taking advantage of people wanting to do the right thing and labelling their bags as '100 per cent degradable'. Avoid these like the plague – it's simply plastic with a chemical additive that speeds up its breakdown into smaller bits. It still stays in the soil or gets washed into oceans.

Anything labelled 'biodegradable' is supposed to break down with the help of living organisms (bacteria etc), but since Australia has no standards governing the use of the term, it can get used and abused (think 'all natural' on a food label). It's best to look for products made from plant-based materials (such as corn, wheat or seaweed) which carry the 'home compostable' green bin logo.

The problem is that even home compostable materials still require the perfect combination of heat, light and oxygen for the microbes to do their thing and break them down. This means it can take around nine months for a bag to break down in a home compost set-up (depending on the time of year, the type of home compost set-up you have etc).

Note that if the material is only labelled 'compostable', it will only break down in a commercial composting facility. This is fine if your local council provides food and organic waste services, but not many do (less than 15 per cent at the time of writing, and another reason to push your council to get this implemented).

And if you do use compostable packaging, never put it in your general waste bin. If these bags and containers end up in landfill they will create more of the methane we are trying to reduce.

The bottom line? Avoid using plastic containers or bags until the technology gets a whole lot better.

Home Compostable

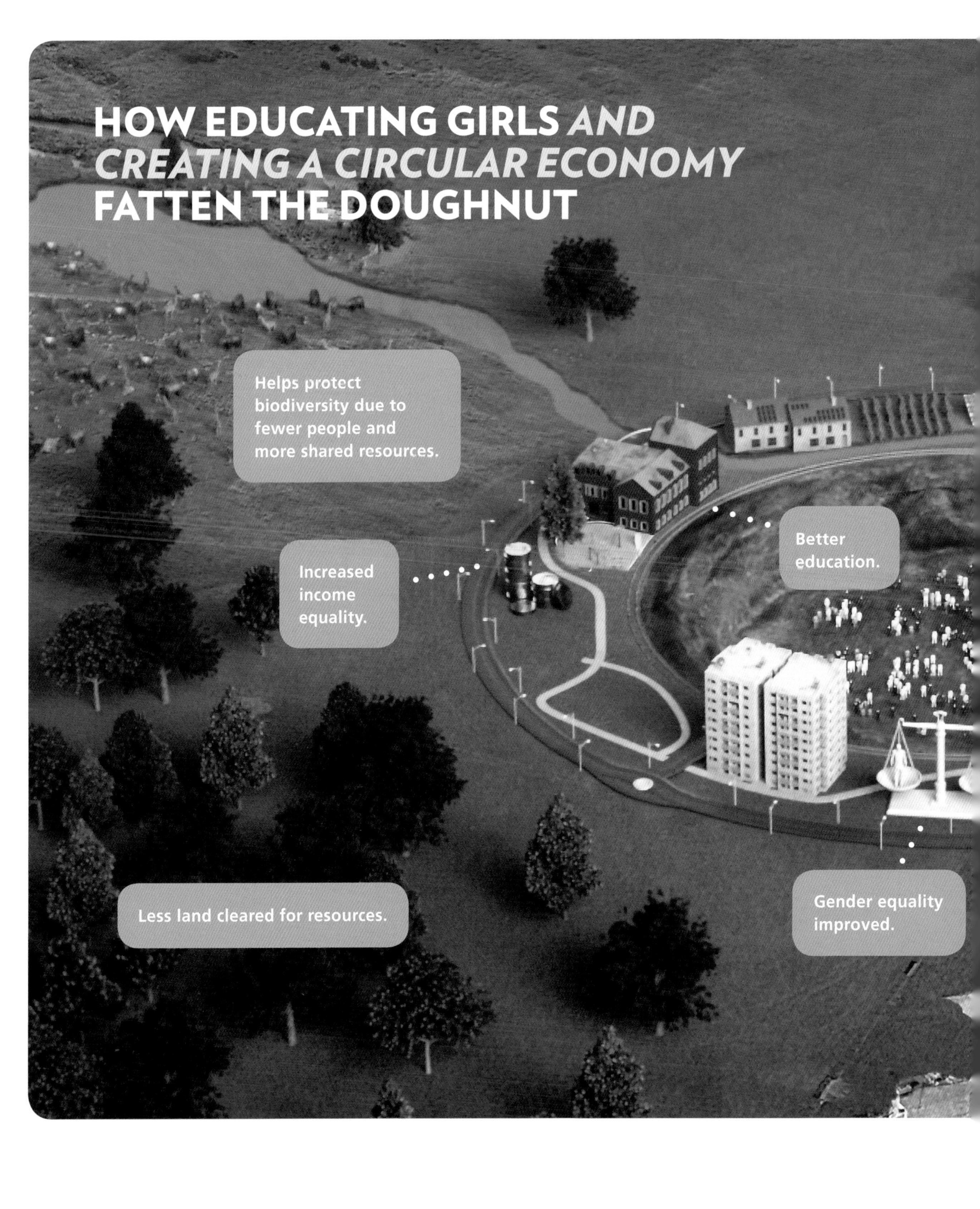
HOW EDUCATING GIRLS *AND CREATING A CIRCULAR ECONOMY* FATTEN THE DOUGHNUT
Helps protect biodiversity due to fewer people and more shared resources.
Increased income equality.
Better education.
Less land cleared for resources.
Gender equality improved.

Better health outcomes.
Less water required by fewer people.
Increased political voice of empowered women.
Climate change reduced due to fewer people and less materials.
Vote

GETTING
TO

2040

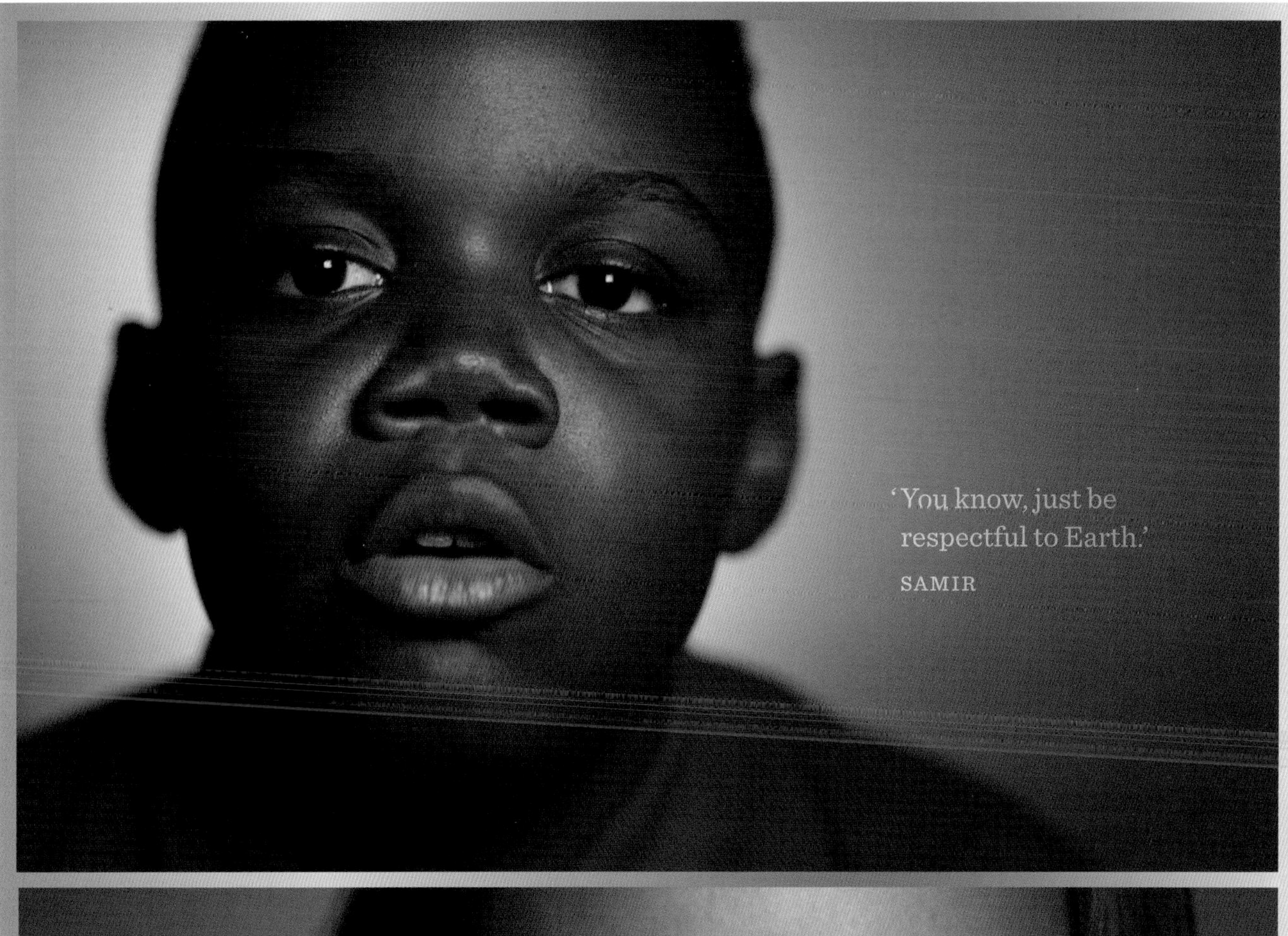
'You know, just be respectful to Earth.'
SAMIR

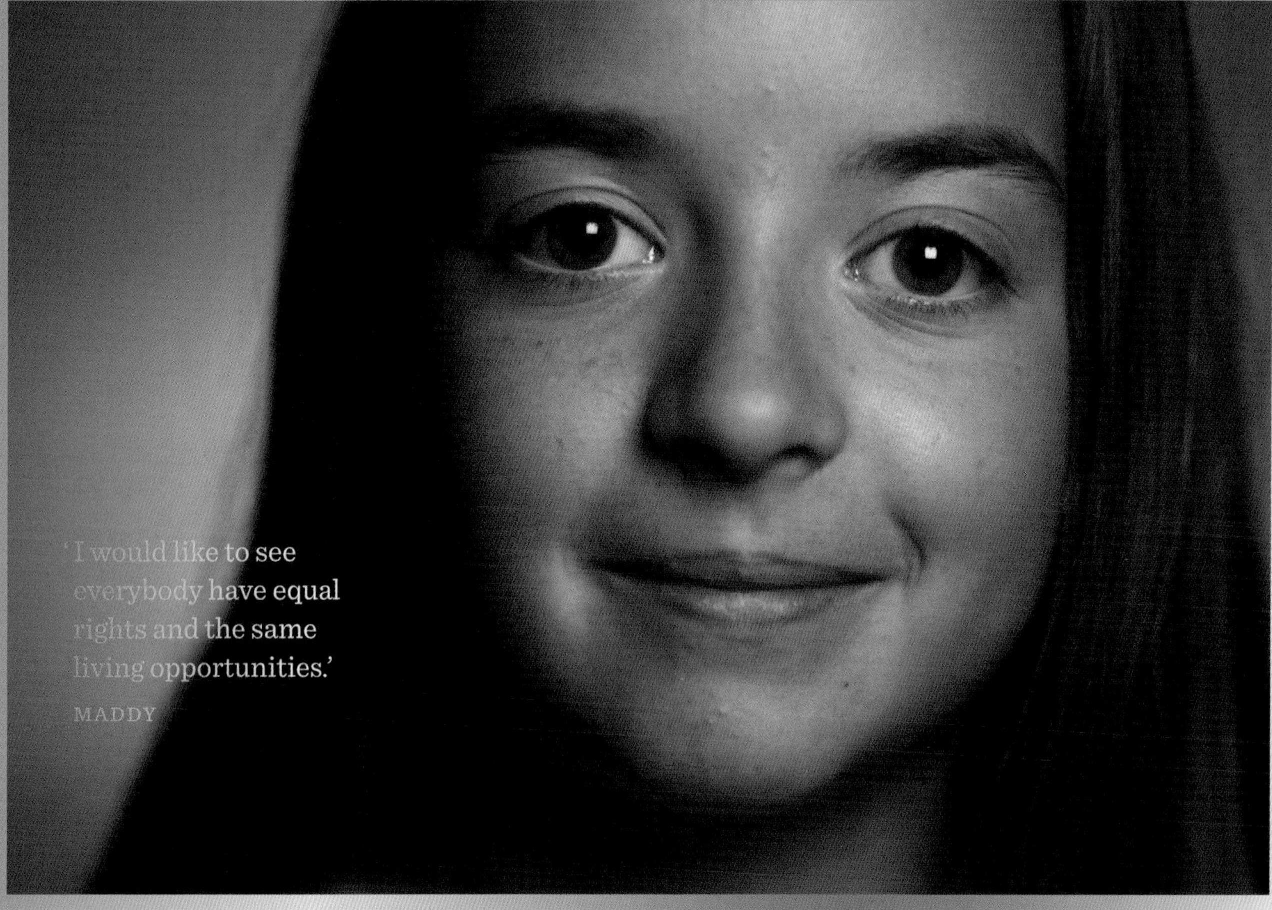
'I would like to see everybody have equal rights and the same living opportunities.'
MADDY

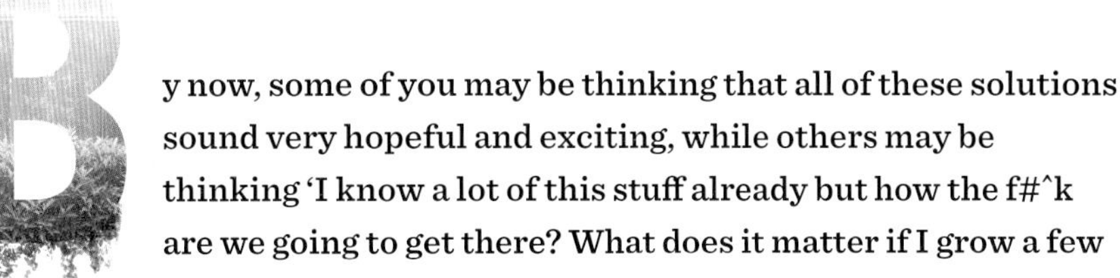

By now, some of you may be thinking that all of these solutions sound very hopeful and exciting, while others may be thinking 'I know a lot of this stuff already but how the f#^k are we going to get there? What does it matter if I grow a few veggies and sing one less song in the shower when one fracking well uses more water than I use in a lifetime of showers?'

While researching how change occurs, a few things jumped out at me. The main point is that change is never linear. We can't propose a simple plan and then expect it to roll out as intended. A heavyweight boxer summed it up best: 'Everyone has a plan 'til they get punched in the mouth.' I grew up believing that society evolves in a steady upward trend. This belief has been knocked about a little over the past few years. Perhaps we need to confront a few more home truths and allow progress to plateau for a bit before it heads in the right direction at a later date? Only time will tell.

I've discovered that an important aspect of change is an 'event'. These are the unforeseen or random moments that can open the door to previously unthinkable changes. This is Rosa Parks refusing to give up her seat to a white passenger who was demanding a seat in the 'coloured' section of a bus after the 'whites-only' section was filled. This brave action or 'event' became the catalyst for a movement and earned Rosa Parks the title of 'the first lady of civil rights'. All the momentum that had been building and bubbling away for a very long time burst through the dam wall with that single 'event' on a bus in Montgomery, Alabama. Apparently, when the former British Prime Minister Harold Macmillan was asked what he feared most in politics, he replied, 'Events, dear boy!'

I have thought about this in regards to climate events. Droughts, wildfires, intense storms and floods have been increasing over the last 20 years, but will there be a tipping point? Will there be a hurricane, typhoon or storm surge that dwarfs the others and inspires global action? It's a horrible thought. But it's worth remembering that if things do begin to fall apart environmentally or economically, then radical ideas can quickly become common sense.

I did find that key movements in recent history share some common themes. Whether it was the abolition of slavery, the human rights movement or the marriage equality campaign, these movements found a way to merge the networks and groups that had been going at it alone for a long time. They made pledges, created viral social campaigns (long before the internet) and managed to change the 'norms' of what is accepted in society. And when our challenge today seems overwhelming, it's worth remembering that every struggle in the past has been fought and won by people with far less wealth and opportunity than we have today. It might be hard to visualise now, but by 2040 it may be less

'The greatest threat to our planet is the belief that someone else will save it.'

ROBERT SWAN

acceptable to wear 'fast fashion', to drive in a car alone, to eat non-regeneratively farmed food, to get your energy from a centralised grid or to drink from anything made of petroleum-based plastic.

While some people (and increasingly news articles) suggest that individual actions don't matter much, they do. Studies show that the important choices we make in our homes (in regards to electricity use and the foods we eat) coupled with our choices around transportation can have a major impact on global emissions. People taking action in their own lives is the best way to build momentum for larger systemic change. We are social animals and we use social cues to guide us. If people are rushing in with buckets of water to put out a fire, others will grab buckets and help. Studies also show that when a home installs solar panels on their roof, the odds of a neighbour installing panels go up, so we will only change our cultural norms and behaviours by making them more visible. This is where those with the knowledge and understanding get to lead the way. The Earth and our children are currently hiring leaders. You in?

While individual actions are important, our current dilemma also presents us with the terrific opportunity to genuinely connect with more people, something which so many of us have been crying out for. Paul Hawken asks us to seriously consider it this way: 'Is global warming happening to you or for you? It's a really interesting question. Because if you think it's happening to you, then first of all you're the object; you're the victim and then you feel like you want to blame someone else. But if you say it's happening *for* me, then you take 100 per cent responsibility. You save that energy for something else – the blame and demonisation – and you say, "Okay, what am I going to do?" It just unleashes the power of innovation, imagination and creativity that sits within all people. It makes a rich, full, extraordinary life. So see where you get lit up, maybe it's girls' education. Wow, I had no idea. Or it can be food, it can be farming, it can be energy, it can be housing, transport. There are so many areas where you actually probably do have influence, where you are working, where you know somebody, where you can effect change in a way that is demonstrable and meaningful to you and others. It's really important to find those places where you feel passionate and it lights you up. As opposed to feeling guilt. Guilt is not going to work for anyone, or the planet.'

Just in case you needed any more of a nudge to kick you into action, political scientists have studied the history of campaigns across the world since 1900 that led to a territory liberation or government overthrow. No campaign failed once it achieved sustained and active participation of just 3.5 per cent of the population.

'Never doubt that a small group
of thoughtful, committed citizens
can change the world; indeed,
it's the only thing that ever has.'

MARGARET MEAD

MR. PRESIDENT
HOW LONG
MUST
WOMEN WAIT
FOR LIBERTY

OTHER PATHWAYS TO A BETTER 2040

I now want to share some of the other suggestions and ideas I encountered on my journey exploring how we might attain a better 2040. Space necessitates that these are not explored in great detail, but it's good to be familiar with them so that we can begin discussing and debating. The 2040 we do reach will only be created by the decisions we make today.

Change our definition of success

Currently, a country's success is measured by what is called GDP (gross domestic product). When you hear 'growth is up 3 per cent' on the nightly news, this refers to GDP. It totals up the value of the goods and services that have been sold in a particular time period (usually a quarter or a year). What it doesn't capture is the environmental damage during the production of those goods and services, the impact on local business when multinational 'superstores' roll into town, or the value of all the unpaid or voluntary work that gets done throughout society (and often keeps it functioning).

Kate Raworth told me about a report from Insure.com in the US that listed every role that a mother plays throughout the day, from childcare worker to cleaner to chauffeur to janitor. If she was paid the going rate for each of these roles, she'd be paid about US$70,000 a year for her work. But right now, her contribution doesn't even show up in national income.

What the GDP doesn't measure remains effectively invisible for policymakers and often for society at large. The suggestion is to make the now-invisible costs, such as environmental damage, more visible because the metrics we define our success by will be the metrics we strive for. Interestingly, 81 per cent of people in Britain now believe that the government's prime objective should be achieving the greatest happiness, not the greatest wealth.

One scene in the film that didn't make it to the final cut was of me watching TV in the year 2040. There was a 'wellness report' on the news that showed someone referring to the GPI (genuine progress indicator). This metric already exists and factors in environmental and social costs alongside the value of goods and services within the economy. The US states of Maryland and Vermont have adopted the GPI, and some European governments are considering doing the same. (This example is of a GPI in Oregon, USA. As you can see, the overall growth would look very different if the environmental costs were factored in.)

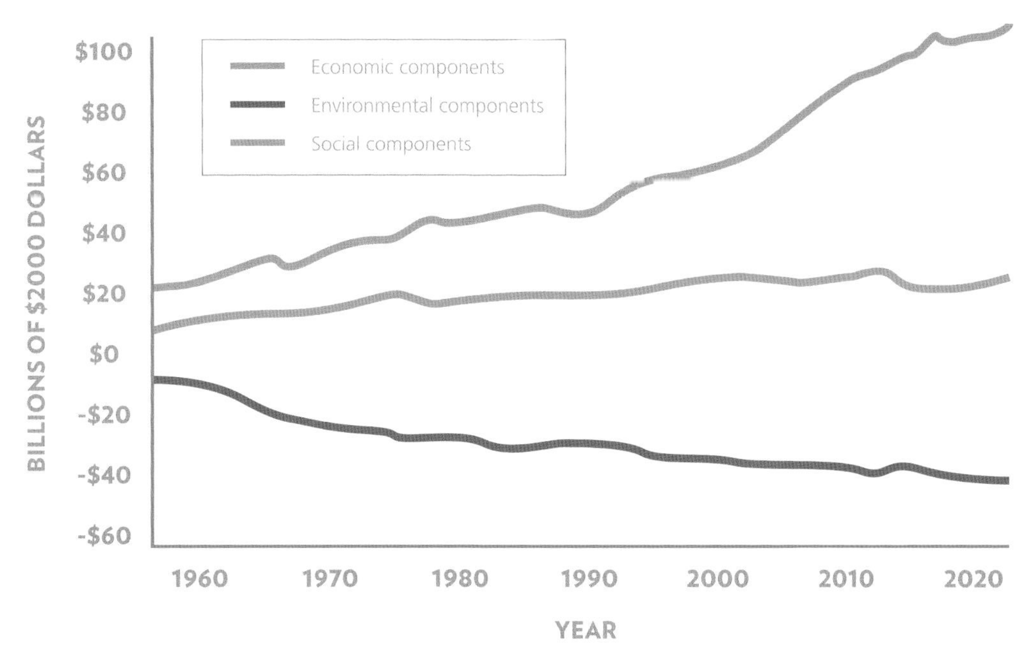

'GDP measures income, but not equality, measures growth, but not destruction, and it ignores values like social cohesion and the environment.'
LORENZO FIORAMONTI
ECONOMIST AND PROFESSOR
OF POLITICAL ECONOMY AT THE
UNIVERSITY OF PRETORIA

Implement trade transparency

Trade treaties are the 'hidden power' of society where decisions that hugely affect the flow of resources and our lives are often kept secret, away from the media. Meetings are often attended solely by a trade or finance minister. Many people have argued that while these crucial decisions are being made, someone at the meeting should also represent our environment and social wellbeing. Perhaps the public could even elect who they want making these decisions?

'Society needs to shape business, rather than business shaping society. This doesn't mean that government is going to run business. That you'll have huge top-down, inefficient structures running business. No. There is absolute value to free enterprise, to creativity, to flexibility. But it's got to be within the bounds determined by society.'

HELENA NORBERG-HODGE

Use 'The Doughnut' as a framework for change

Using Kate Raworth's doughnut model, we need to be living within the safe space of the dough. Unfortunately, we have already eroded the doughnut's outer boundary with climate change, biodiversity loss, ocean acidification, and nitrogen and phosphorous loading. We have also damaged the inner boundary by falling short on health, water, food, education and political voice. This is effectively reducing the safe space in which we can thrive. Imagine if we used these boundaries to frame the important decisions we make moving forward? There would still be inequalities, and there would still be growth, but it would need to be within the limits of our own and the planet's wellbeing.

Help farmers transition

The hope of many people I interviewed is that agricultural solutions will receive the same attention that renewables do in the media, with philanthropists and investors helping farmers transition to more regenerative practices. Investment opportunities are increasing in this area.

A carbon tax could also help the transition process. The money raised from the tax could pay farmers to put the carbon back into their soils, improving the quality of our food and retaining precious water. (It's worth noting that Australian farmers receive fewer government subsidies than almost every one of the 52 OECD nations. With more and more farmers under pressure, the other option is that we stop subsidising fossil fuels in Australia and send some of that money to farmers to put carbon in the soil and transition to regenerative practices.) While we're waiting for that, check out **Carbon8** (carbon8.raisely.com) and see how you can support a farmer to put carbon into their soils for the price of just two lattes a month.

Other suggestions around a carbon tax (there are numerous) involve using the money raised to pay citizens. The public would receive a dividend as we cleaned up our atmosphere.

Between 70 and 80 per cent of the food in the world comes from smallholders and the rest from industrial agriculture. In 2016 the US Department of Agriculture spent the best part of $20 billion subsidising Big Ag. Even a slight shift in subsidies to support smaller producers could have a huge impact. Many of the larger companies grow food for animals (which, as we know, they don't need – they can eat crop residue, grass or food waste) or they grow sugar cane (which we definitely don't need!).

It would also help if the World Trade Organization changed its rules so poorer nations could be more competitive. In some cases, rules stipulate that food produced in wealthier nations can be subsidised by their governments but food grown in poorer nations cannot. This means the poorer nations cannot compete with the cheaper imports from the likes of the US or Europe, and so their economy and wellbeing suffers.

Use your purchasing power

An economist told me that, despite increasing our call for larger structural changes, every purchase we make as individuals has an impact. We dictate the shape of the economy with what we buy. If no one buys DVDs, they go out of business. If no one purchases takeaway coffee cups, the same fate awaits. We can send a clear signal to companies and dictate what they do or don't sell. So being conscious about your spending is important.

'The conscious and intelligent manipulation of organized habits and opinions of the masses is an important element in democratic society. Those who manipulate this unseen mechanism of society constitute an invisible government which is the true ruling power of our country. We are governed, our minds are moulded, our tastes formed, our ideas suggested, largely by men we have never heard of.'

EDWARD BERNAYS, 1928

Curb advertising

Globally, more than $600 billion is spent on advertising each year. Americans are now purchasing double what they did in 1950, and the average child in the US watches 40,000 commercials a year.

Paul Mazur, the business partner of Edward Bernays (the 'father of public relations') said in 1927: 'We must shift America from a needs to a desires culture. People must be trained to desire, to want new things even before the old have been entirely consumed. We must shape a new mentality.'

Sao Paulo in Brazil, a city of 20 million, banned advertising in its city. Have a look at the images online. It has been hugely controversial but billboards have been taken down and people are noticing historic buildings for the first time. This practice has also been adopted in other cities around the world. It is an interesting step towards freeing ourselves from consumption.

Reduce debt

A large reason economies have to constantly grow is because of debt. And since debt comes with interest, it grows exponentially. If you start a business, you may take out a loan, but you'll have to pay that loan back with compound interest – thus you'll need to grow, and thus more resources will need to be used or created. The same rule applies for a business, country or individual. The anthropologist and author Jason Hickel has called for some kind of reduction in debt. 'Creditors like the banks would lose out but isn't this a sacrifice worth making? They've had their fun.' This sounds like a bold suggestion, but when you understand how banks make their money (by literally creating it out of thin air), reining them in doesn't seem like such a bad idea. Some countries in the developing world have already paid off their debts from 30 or 40 years ago a few times over, but are still paying compound interest on their loans, which is impacting their ability to spend money on social services or to mitigate against increasingly hostile climate change (these loans were often made to corrupt dictators, and yet the impacts are now being felt by locals decades later).

Jason Hickel also suggests making banks keep a bigger reserve behind the loans they make (fractional reserve lending), which is currently only around 10 per cent. This means 90 per cent of the money in circulation is debt and as a result, growth is needed to repay loans with compound interest. He also suggests that we could 'abolish debt-based currency altogether and so instead of letting commercial banks create our money, we bring it back to a secure not-for-profit or the state but under a truly transparent, democratic process led by an independent agency. This agency becomes responsible for lending money. The banks could lend money but only against a reserve of 100 per cent that they have.' This idea was endorsed by two IMF economists in 2012 who suggested it would make the economy far more stable. Check out the positive money campaign in the UK (**positivemoney.org**).

I repeat, if things begin to fall apart environmentally or economically, then radical ideas can quickly became common sense.

Clean up social media

A few experts told me that this is a larger problem than we think. Because so many people use these platforms now for opinion and discussion, if the platform is being manipulated then so are people's opinions. Some search engine algorithms, for example, can often offer up extremes. Type in 'vegetarian' on YouTube and you'll get recommendations for veganism; search for 'running' and you'll get recommendations for marathons; even searching for 'Islam' can end up in a world of ISIS.

As we learnt from the recent Cambridge Analytica story, both the 2016 US election and the UK Brexit vote were heavily influenced by online algorithms that created fake people ('bots'). This can give a sense of there being more support for a political campaign than there actually is. A recent study showed that around 66 per cent of the Make America Great Again hashtags (#MAGA) during the 2016 US election were coming from bots. These bots can often be deliberately aggressive and provocative, too. Check out the 'botcheck' website to see if you are in fact arguing with a robot (many people have and the current US President has even re-tweeted one).

Social media can also give the impression that we are more divided than we actually are. A 2018 US study was conducted, based on a nationally representative poll with 8000 respondents, 30 one-hour interviews, and six focus groups conducted over a year. They found that the most loud and vocal people were activists on the left (8 per cent) and devoted conservatives (6 per cent). In the middle was what they called an 'exhausted majority' who were fed up with the polarisation.

One person I spoke to hoped for a future where we were not subject to the algorithms of Facebook, Twitter or YouTube, but were instead subject to 'public interest algorithms'. These would favour society instead of advertising or investor interests. We could choose what content gets prioritised, we could choose to see more content that doesn't fit within our bubble and we could remove or filter sensationalist stories. These algorithms could promote a greater level of understanding online.

Cleaning up mainstream media

When I released *That Sugar Film* in the USA, I did some interviews on a few major television networks. Before we went 'live', I was often given a talking to about what I could and couldn't say on the show. 'Sugar can't be linked to type 2 diabetes or fatty liver disease' was the general brief. It struck me how controlled the messaging can be on these stations, many of which are receiving millions of dollars in funding from advertisers.

In the US, just six corporations (News Corp, Comcast, Disney, Viacom, Time Warner and CBS) control 90 per cent of what is seen, heard or read. In 1983 it was 50 corporations. In Australia we are headed in a similar direction with a more concentrated media landscape. This is not an ideal recipe for democracy. How can we expect to hear stories about Earth Overshoot Day or climate change if they directly challenge how many of the network advertisers are currently operating? Perhaps it is time for a more decentralised 'Public Trust Network'. Would you watch PTN?

'If you added up the biggest co-operatives in the world, they'd actually equal the seventh largest economy in the world. That's almost a hidden story, because people don't realise how much this is coming through and how much new kinds of business and enterprise are bubbling up.'
KATE RAWORTH

Decentralise

Blockchain technologies (if run on renewables, as they currently chew through energy) could allow more secure, decentralised forms of ownership because the technology is on our side for the first time (think of the distributive solar technology with the Bangladesh microgrids – no central utility or ownership required). We need to seize these technologies and come up with new business models, new financing models, peer-to-peer. Perhaps we could see an even bigger rise of co-operatives where workers collectively own their company. In 2012, co-operatives earned $2 trillion in revenue globally.

New Zealand has recently had huge success with crowdfunding equity sites like pledgeme.co.nz, which has raised $30 million in just four years (this practice is now legal in Australia). A great example comes from Dunedin where a Cadbury chocolate factory closed down. The workers got together and crowdfunded the equity to start their own chocolate company. They raised $2 million in record time and now all have a share in the company, which distributes the revenue more evenly. There are increasing examples of this model. Why not solve some of our largest problems while increasing income equality at the same time?

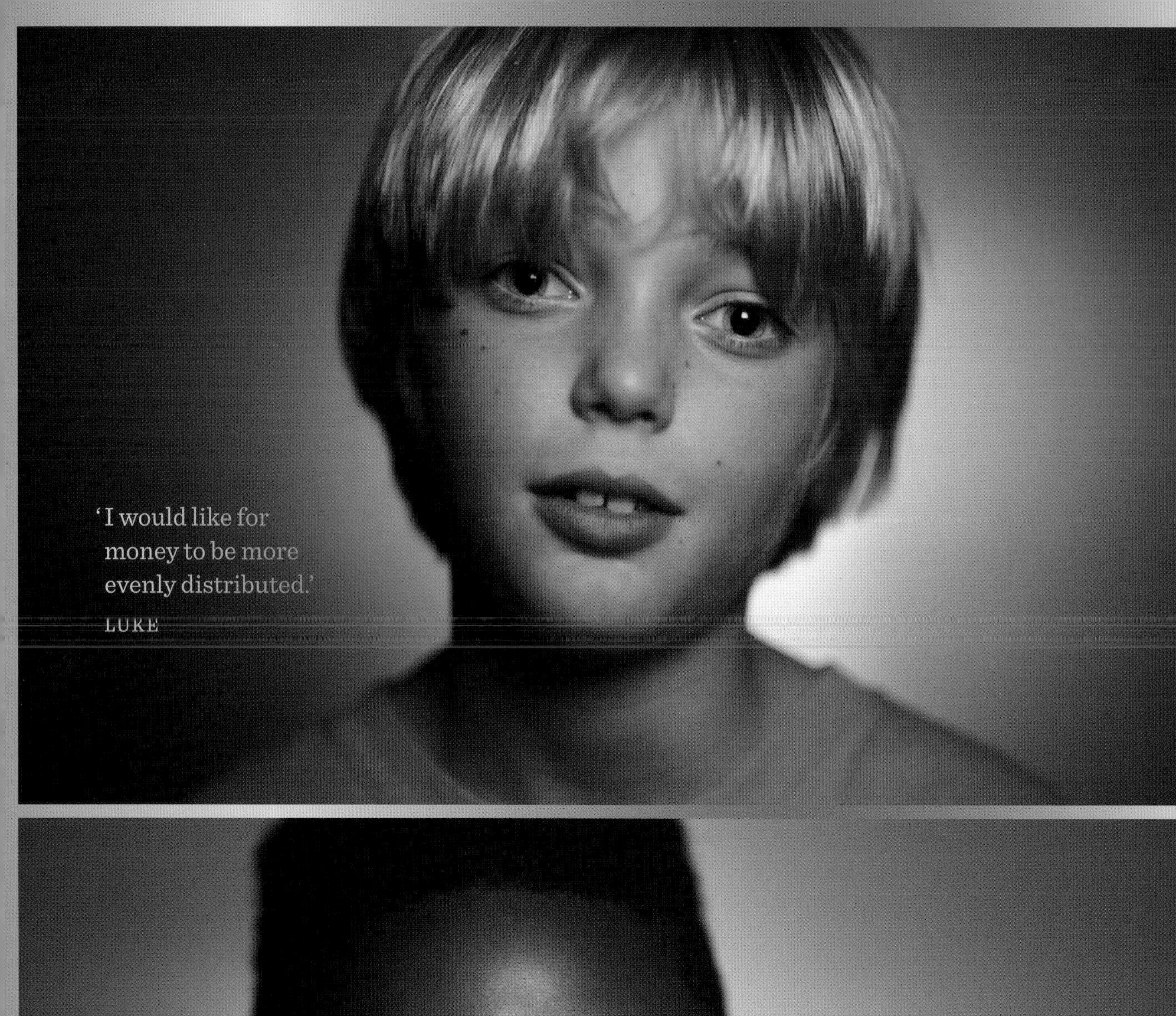
'I would like for money to be more evenly distributed.'
LUKE

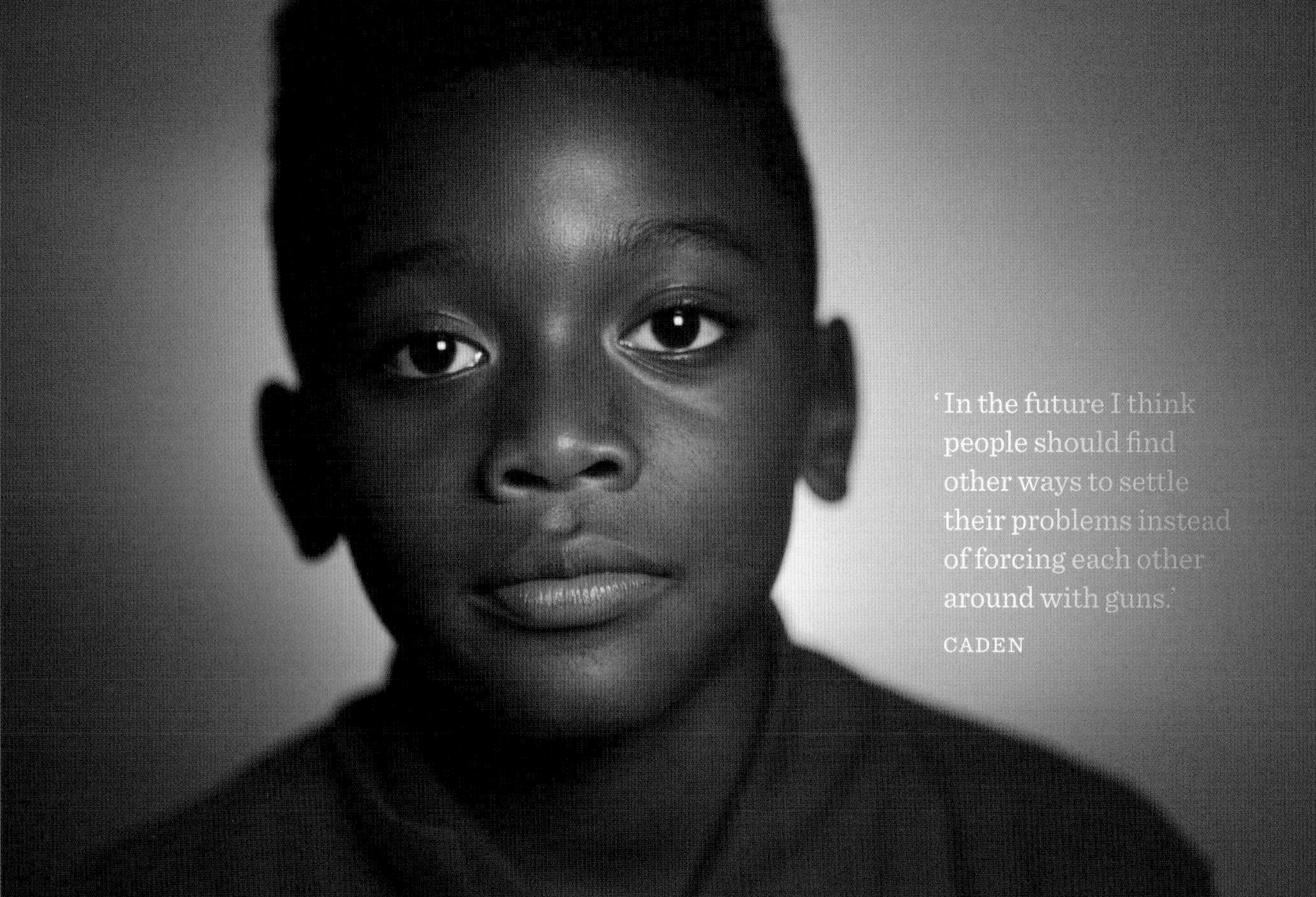
'In the future I think people should find other ways to settle their problems instead of forcing each other around with guns.'
CADEN

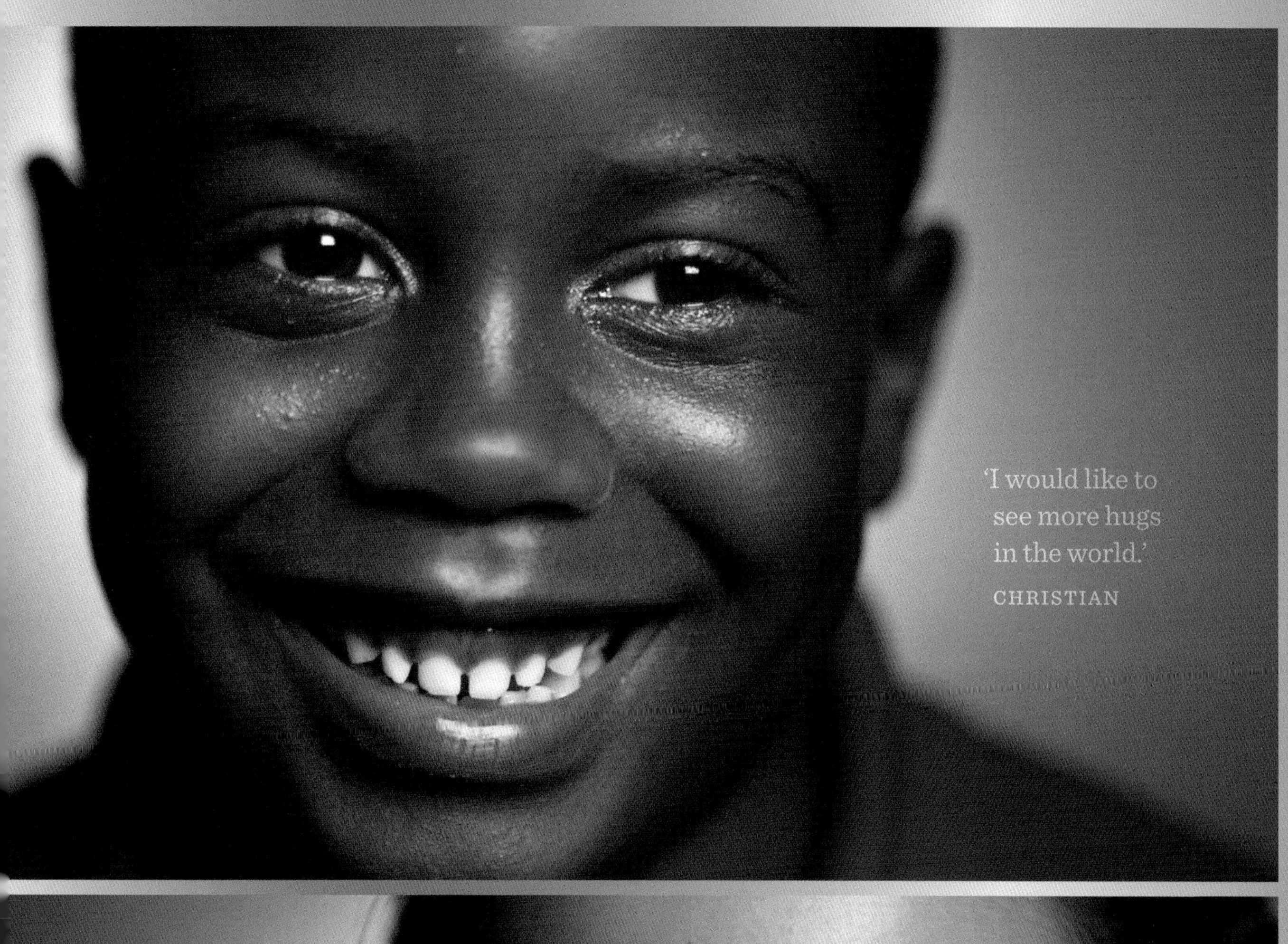

'I would like to see more hugs in the world.'

CHRISTIAN

'Well I'd like it to be human instinct to just look after the world and to care for the world.'

STELLA

Strengthen democracy

It's fair to say that politicians are not highly regarded in many parts of the world at the moment. As a result, many political experts, economists and think tankers are searching for ways to improve our democracy. Whenever I asked how we might achieve this, 'get money out of politics' was the most common response. In the US, a whopping $3.3 billion per year is spent lobbying politicians, and for every $1 that public interest groups spend lobbying, corporations spend $34. Consider that in 1972 when a magnate/philanthropist gave the Richard Nixon campaign US$2 million (about US$11 million today), public outrage led to reforms in campaign financing; in contrast, the war chest accumulated by the Koch brothers and their friends for the Republican party in the 2016 election was around US$889 million.

Obviously structural issues need to be addressed, but a sliver of light comes from a practice called 'participatory democracy', which is gaining traction in many parts of the world. It started in Brazil in the late 1980s and involves residents coming together to democratically decide how their government's budget should be allocated. It gives value and weight to people's views, plus it helps shift finances to a local level where they can be best utilised. Most countries that practise participatory democracy have seen the prioritisation of social services and the environment.

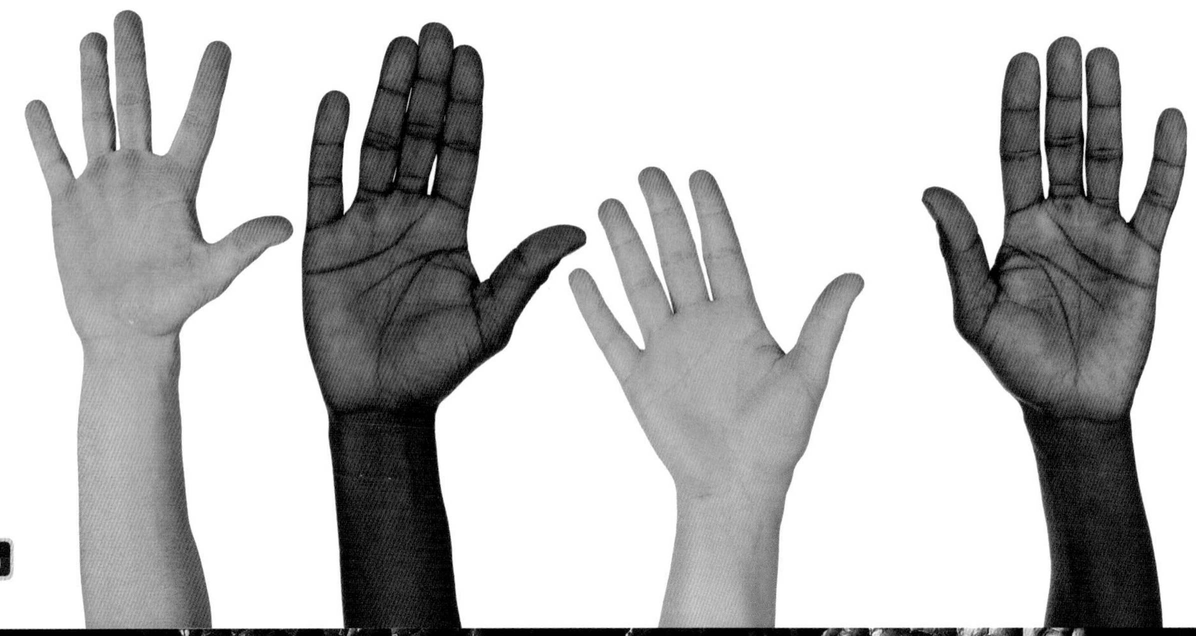

Consider emergency mobilisation

Some people believe we need urgent climate mobilisation to reach a better 2040, not unlike when Franklin D Roosevelt (FDR) put out the call for help at the start of the Second World War. Families planted 50 million 'Victory Gardens' that supplied 40 per cent of America's vegetables during the war, and factories stopped making vacuum cleaners and washing machines and instead made machinery for the war. That period saw great innovation and huge technological breakthroughs, and 10 per cent of the US population relocated, often across state lines, in order to find a 'war' job. They were driven by a purpose and to serve their country. The combination of full employment and progressive taxation during this time caused income inequality to plummet. Profit-seeking behaviour was either subordinated or channelled towards the mission to win the war. This would send shivers up the spine of any anti-government, climate-denying Libertarian. But I do wonder: if we don't take the necessary action soon, could something like this be non-negotiable in the future? There is building momentum in the US right now for a 'Green New Deal', which proposes progressive tax rates, the removal of fossil-fuel subsidies and the creation of millions of jobs to remove carbon from the atmosphere. As I write, there is still ambiguity around how 'democratic' the proposal actually is, but this deal could receive a lot more attention and scrutiny in the near future.

Will you have a part in Victory?

WRITE TO THE NATIONAL WAR GARDEN COMMISSION — WASHINGTON, D.C. for free books on gardening, canning & drying.

"Every Garden a Munition Plant"

When the banks needed to be bailed out after the crash in 2008, at least $8 trillion was found (some estimate $16 trillion, even $29 trillion) to tidy up the mess because the banks were 'too big to fail'. You'd hope that the fate of the planet and humanity would also be 'too big to fail'. Considering that there's around $18 trillion hidden in offshore tax havens, perhaps these havens could be shut down and jail terms could be pardoned if the perpetrators agree to give a predetermined figure, say 50 per cent of their earnings, to help regenerate the planet? Fair deal?

One last time: if things begin to fall apart environmentally or economically, then radical ideas can quickly become common sense.

A NEW WAY

While the solutions in the last few pages could play an important role in creating a better 2040, many of them are levers inside an existing paradigm. The author and systems thinker Donella Meadows believed that real change in a system can occur when people are shown a new way. This is the primary reason I made the film for our daughter.

Since the scientific revolution of the 1600s, we have viewed our world in a highly mechanistic way. The earth is here to serve us, and we are here to conquer and extract things from it. But, increasingly, biology is reminding us of an older story that our ancestors knew to be true. It is a story of interconnectedness, a story of dazzling co-creation between all forms of life, whether in the depths of the soil, in the conversations between trees, or how the critters in our guts meticulously work with our bodies to maximise the foods we deliver them. What is fundamental to our survival is not only a re-enchantment with the natural world, but a deep understanding that any actions we inflict upon it are the same actions we inflict upon ourselves.

For too long, scientists alone have been left to communicate to us what is happening in our world, and while their message is crucial, their language is specific to their fields. We need to bring back the love, the poetry, the wonder, the reverence and the meaning to our view of our planet. This is a task to be shared across many disciplines and involves finding a new language to describe our journey. Words and phrases like 'negative emissions', '2 degrees warming' or 'anthropogenic' serve an informative role but rarely pique the interest of our hearts. We need to come together and decide what new words will ignite and unite us, what words will evoke meaning and will spur us into more care and action. (I will offer up replacing 'the environment' with 'our living planet', or replacing 'deforestation' with 'destroying living systems' as a start.)

I had a powerful lesson in the importance of meaning a few days ago. Our daughter's pet rabbit passed away at the vet and, as I watched it take its last breath, I was surprised at how emotional I felt. As a child I'd often gone rabbit shooting with family friends. Rabbits were considered pests on South Australian farms and I remember feeling a nervous excitement when the spotlight hit a rabbit frozen beside some spinifex. But here I was shedding a small tear at the death of our daughter's rabbit because this was a rabbit with meaning. It had been a part of our lives for over a year, it had run amok in our house, chewed our power cords, been regularly cuddled to within an inch of its life and had stamped its personality on our family. It was the same creature as the one awaiting death in the spotlight yet it was completely different.

What if, collectively, we gave meaning to our planet? What if we saw it as a precious and delicate home? Would we sit by and watch as more and more of its creatures became extinct? Would we permit the degradation of the teeming life in our soils that play such a pivotal role in our health?

What if the single most important thing we could do to save our planet was to teach our children to love it, to value it and to give it meaning?

The Planet Ark Environmental Foundation recently conducted a survey about children's contact with nature. They highlighted a dramatic shift from outdoor play to indoor play in a single generation. Almost one in three kids has never climbed a tree, or planted or cared for a tree, and for every hour a child spends outdoors, seven hours are spent in front of a screen.

**'Only if we understand, will we care.
Only if we care, will we help.
Only if we help, shall all be saved.'**

JANE GOODALL

WE HAVE SOME *WORK TO DO*

To respect and love our planet, it is equally important that we show ourselves the same respect. As I mention in the film, it is very hard to avoid hypocrisy at the moment as the majority of things we do harm the planet in some way. But if we are to reach a better 2040, I think it's important that we try to avoid feelings of guilt. The psychologist Renee Lertzman told me that this emotion can prevent us from taking action.

What we are currently facing is a transition phase which, as we all know, can be very clunky. Think about how drawn out, awkward or painful some of your breakups have been! We are essentially trying to move on from a partner who was fast-moving, exciting and pretty addictive, and we're now aiming for a deeper connection that has marriage material written all over it.

But, as with any breakup, there are many ways to deal with it. Some of mine were handled with grace (usually not by me), while others were downright nasty. The nasty ones still evoke a small 'twang' inside me. In my experience, approaching any kind of change with too much hostility or anger is counterproductive and usually leaves a bitter aftertaste.

I raise this as a metaphor for the environmental transition we will have to embark on. I personally disagree with the 'fight climate change' narrative that often appears in our media. To 'fight' our environment comes from the same mechanistic approach that got us into this mess in the first place. We are a part of our environment so why would we want to fight ourselves?

Remember the natural carbon cycles of our planet (page 99)? The way out of our predicament is to work with these natural cycles, not against them. This would prevent the further release of stored fossil fuels while returning atmospheric carbon to a safe and familiar home, where it will be welcomed by our soils and plants who eagerly await its arrival and bear numerous gifts.

Some people might find the thought of this new reality a little too utopian. On the contrary, I think it's utopian to believe that we can keep extracting resources from the earth in the way that we have been without any consequences.

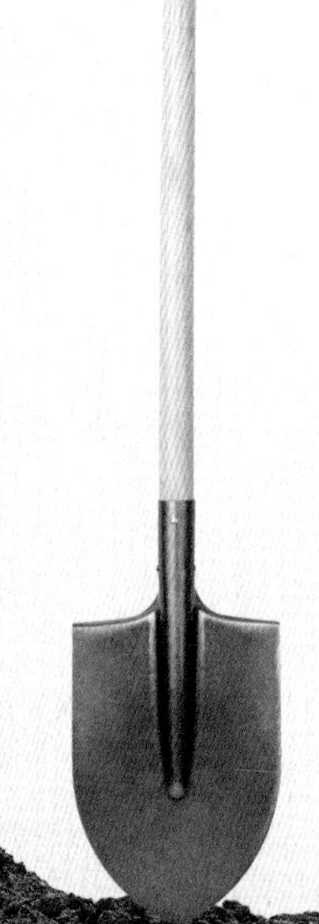

Other great transitions in history have been similarly labelled. When attempting to outlaw slavery, the abolitionists were often derided for their utopian vision. At the time, many believed that the economy would collapse without slaves, that they were an essential cog in the wheel of civilisation. If you had mentioned ending slavery 100 years before its actual demise, nine out of 10 people on the street would have thought you were bonkers.

Even those calling for women to get the vote around 100 years ago were labelled as utopian. 'Sensible and responsible women do not want to vote,' said Grover Cleveland, former US President. 'The relative positions to be assumed by man and woman in the working out of our civilization were assigned long ago by a higher intelligence than ours.'

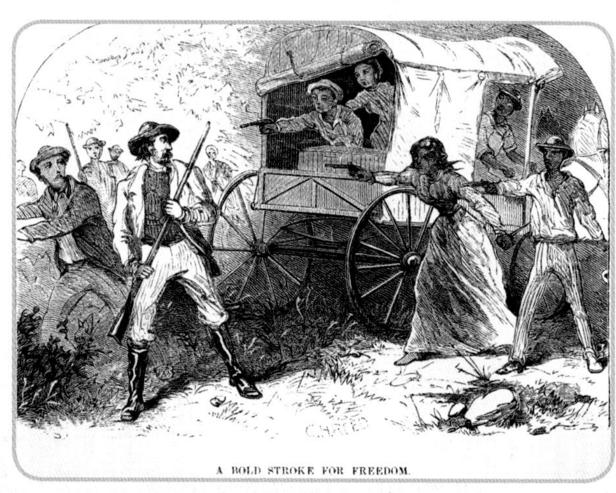

(Clockwise from top left): Frederick Douglass (1818–1895), US social reformer, abolitionist and writer; suffragettes in Washington, D.C. and New York (1910s); illustration from William Still's *The Underground Railroad Records* (1872).

Here are some other classics:

'This "telephone" has too many shortcomings to be seriously considered as a means of communication.'

WILLIAM ORTON, WESTERN UNION PRESIDENT, 1876

'No online database will ever replace your daily newspaper.'

CLIFFORD STOLL, US ASTRONOMER, 1995

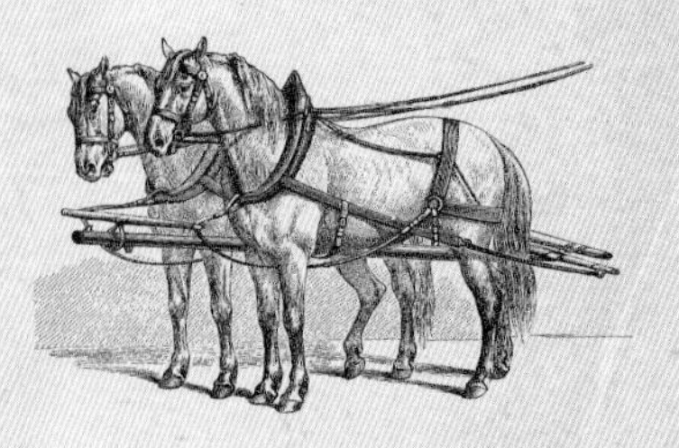

'I don't believe the introduction of motor-cars will affect the riding of horses.'

MR SCOTT-MONTAGUE, UK MP, 1903

'Coal is going to be an important part of our energy mix, there's no question about that, for many, many, many decades to come.'

MALCOLM TURNBULL, AUSTRALIAN PRIME MINISTER, 2016

Striving for a better 2040 should be embraced and celebrated. It's the world our children will inhabit. Getting there should not be seen as a sacrifice but as an opportunity to create a better home for every living thing. A health scare is really just feedback from our body that provides us with the opportunity to change and live a healthier life. Well, we are also receiving feedback from our planet, and it is also offering us the opportunity to change and live a healthier life. A life of cheaper electricity, healthier food, thriving ecosystems, cleaner transport and connected communities. It certainly won't be perfect, but it's better than our current trajectory, which sees a future of an elite minority living behind gated walls while the rest of society scrambles for resources on an increasingly uninhabitable planet. Sadly some of our societies are already there. None of us want this future for our children and none of us are as selfish as we are often led to believe. Countless studies show that, despite possessing selfish traits, we are first and foremost kind and empathetic beings who crave connection and community. (There's a reason 'solitary confinement' is a form of punishment.) What is now laid out before us is the opportunity to remain as brilliant and colourful individuals, but to come together and tell a new story about our future. Because if we don't come up with a vision of the future together, then we will simply become a part of someone else's vision.

The reason I planted a tree with my family at the beginning and end of the film is that it provides the best metaphor for what we need to do. In essence, we are simply planting a seed. But that seed will grow and become a tree one day. A tree that may benefit the soil, provide a sanctuary to an array of life, or even be used to build a beautiful home for someone. You are unlikely to know the fate of that tree because it will probably outlive you. But so will the simple actions that you take today to help our planet. They may do more for future generations than you will ever understand.

'Unless someone like you cares a whole awful lot, nothing is going to get better. It's not.'

DR. SEUSS, *THE LORAX*

To learn how you can get more involved in the Regeneration please head to whatsyour2040.com and 'Activate Your Plan'.

RECI

The following recipes are all vegetarian and contain many ingredients that either benefit the soil or sequester large amounts of carbon from the atmosphere. I am not advocating for a vegetarian or vegan diet, but by choosing to eat a few of these meals a week instead of conventionally farmed animal protein, we can reduce our impact on the planet. Please note that some regeneratively farmed animal protein can also sequester large amounts of carbon and improve soil health.

PES

PERENNIAL FOODS THAT SEQUESTER CARBON AND BUILD SOIL HEALTH

Eating these foods will have the biggest impact if they are sourced as locally as possible, nothing is wasted – and no rainforests are cut down in order to grow them.

1_BLUEBERRIES
2_DATES
3_BANANAS
4_MANGO
5_OLIVES
6_PEACH
7_RASPBERRIES
8_BLACKBERRIES

FRUIT

9_STRAWBERRIES
10_ORANGE
11_LIME
12_APPLE
13_APRICOTS
14_PEAR
15_NECTARINE

FRUIT

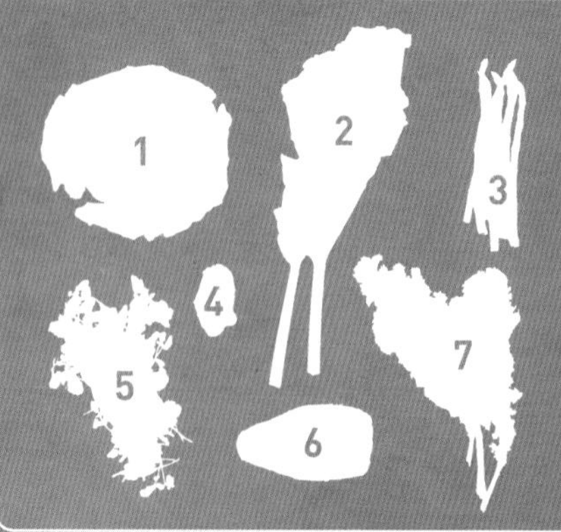

1_RADICCHIO
2_RHUBARB
3_ASPARAGUS
4_JERUSALEM ARTICHOKE
5_WATERCRESS
6_TARO
7_KALE

VEGGIES

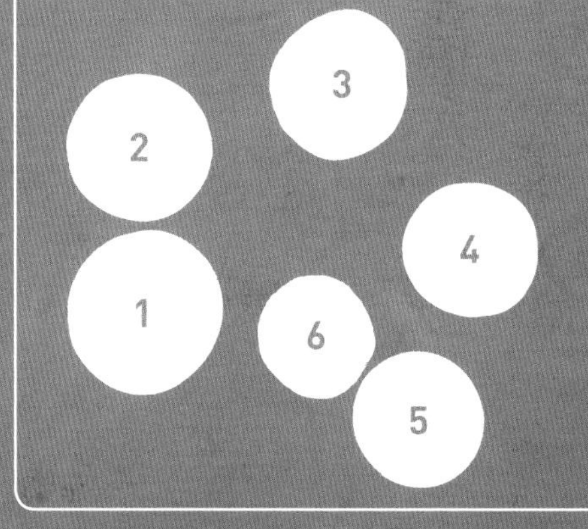

1_COCONUT
2_HAZELNUTS
3_ALMONDS
4_PISTACHIOS
5_GOJI BERRIES
6_RAW CACAO POWDER

NUTS, etc

SEA VEGETABLES

Sea vegetables sequester large amounts of carbon from the atmosphere and can help to alkalise the ocean. They are also pretty darn good for our own health.

1_WAKAME
2_NORI
3_DULSE FLAKES
4_DULSE

5_MIXED SEAWEED

6_ARAME

7_TASMANIAN KOMBU

BREAK

FAST

SUMMER CHIA OATS

Sometimes you need a break from eggs, and so does the planet. Oats, especially organic ones, are a great choice as their production results in fewer emissions than animal products. Choosing Australian-grown chia seeds over imports will reduce food miles too.

SERVES 4

PREP: 15 MINUTES, PLUS 15 MINUTES STANDING TIME

2 cups whole rolled oats

⅓ cup chia seeds

2 tablespoons linseeds

2 tablespoons sunflower seeds

2 teaspoons ground ginger

1⅓ cups boiling water

½ cup macadamias, toasted and chopped

1 cup milk of your choice (try oat, almond, hazelnut or coconut)

1 small mango, peeled, seed removed and flesh thinly sliced

2 heavy passionfruit, seeds and juice scraped

⅛ small pineapple, peeled and sliced

2 tablespoons pistachio kernels, lightly toasted and roughly chopped

mixed berries, to serve

coconut yoghurt (optional), to serve

Place the oats, chia, linseeds, sunflower seeds, ginger, boiling water and half the macadamias in a large heatproof bowl and stir until well combined. Set aside for 3 minutes or until the water has been absorbed.

Stir in the milk and stand for a further 8–10 minutes or until the milk has been absorbed and the mixture is thick and creamy (similar to a porridge consistency). Add a little more water if needed.

Divide the oat mixture evenly among bowls and top with the mango, passionfruit and pineapple.

Serve sprinkled with the pistachio kernels and remaining macadamias, plus the berries and coconut yoghurt, if you like.

CAULIFLOWER TOASTS

Try swapping your regular bread for this nutrient-dense cauli toast and I promise you won't regret it – it is well worth the extra effort. To make things a bit easier, you can prep and steam the cauli the night before (just store it in an airtight container in the fridge) and cook up the toasts and rainbow chard mixture in the morning.

SERVES 2–4 (DEPENDING ON WHETHER YOU HAVE 1 OR 2 SLICES OF TOAST PER SERVE)

PREP: 25 MINUTES, PLUS COOLING TIME

COOKING: 35 MINUTES

450 g piece cauliflower

1 free-range egg

30 g parmesan, finely grated

2 teaspoons sweet paprika

sea salt and freshly ground black pepper

avocado oil, for cooking

1 small leek, white part thinly sliced

4 stems rainbow chard, stems thinly sliced and leaves torn

300 g cherry tomatoes, sliced

⅓ cup herbs (such as dill, flat-leaf parsley and/or basil), torn

lemon wedges, to serve (optional)

Cut away the florets from the cauliflower, then roughly cut them into 3 cm pieces. Chop the stem into 1 cm pieces and tear any leaves in half. Place in a steamer basket set over a saucepan of boiling water over high heat. Steam for 5 minutes or until just tender. Transfer to a colander to drain and cool to room temperature.

Place the cauliflower in a food processor and, using the pulse button, process in bursts until finely chopped. Add the egg, parmesan and paprika and blend together for 20 seconds until well combined. Season well. Divide the mixture evenly into four.

Heat a large cast-iron frying pan over medium–high heat. Pour in some oil and, using two portions of the cauliflower mixture, press and shape them in the pan into two separate 1 cm thick rectangles roughly 10 cm × 7 cm. Cook for 3–4 minutes until set firm on the underside and golden, then carefully turn and cook for a further 3–4 minutes or until golden. Transfer to a plate and cover with a clean tea towel to keep warm. Repeat with more oil and the remaining cauliflower mixture.

Increase the heat to high and add more oil to the same pan. Once hot, add the leek and rainbow chard stems. Cook, stirring constantly, for 3 minutes or until softened and light golden. Add the rainbow chard leaves and tomato, and toss gently for 1 minute or until the leaves are only just starting to wilt and the tomato is just warmed through. Remove the pan from the heat, toss through the herbs and season well.

Divide the cauliflower toast among plates and spoon over the rainbow chard mixture. Serve with lemon wedges, if you like.

WAYS TO USE YOUR LEFTOVER JUICING PULP

Get creative with your pulp to save money and reduce food waste.

ADD IT BACK INTO YOUR JUICE!

The pulp (skin and seeds included) contains so much goodness – really fantastic for gut health.

LAYER IT ON MUESLI

Layer fruit pulp with your favourite muesli and yoghurt in chilled glasses and serve as a breakfast parfait.

MAKE BURGERS

Use leftover vegetable pulp to make veggie and bean burgers.

FREEZE IT

Freeze fruit or vegetable pulp in ice-cube trays. Add them to your water bottle as an on-the-go flavour boost, or plop them into chilled glasses of sparkling water.

BAKE IT

Try incorporating fruit or vegetable pulp (or a combination of both) when baking cakes, muffins, loaves and brownies. See the Fruit and nut crumble opposite, and the Carrot, almond and lemon loaf on page 234.

FRUIT and NUTS CRUMBLE

SERVES 6

250 g strawberries, hulled and halved
1 red apple, very thinly sliced
1 overripe pear, very thinly sliced
1 tablespoon maple syrup
½ cup leftover fruit juice pulp
¾ cup whole rolled oats
1 cup chopped mixed nuts
2 teaspoons ground cinnamon
2 tablespoons macadamia oil
coconut yoghurt, to serve (optional)

Preheat the oven to 200°C (180°C fan-forced). Toss together the fruit and maple syrup in an ovenproof dish. Bake for 15 minutes or until softened. Meanwhile, combine the remaining ingredients. Remove the fruit from the oven and top evenly with the pulp and oat mixture. Return the crumble to the oven for 15–18 minutes or until golden. Serve warm with coconut yoghurt, if you like.

CARROT, ALMOND and LEMON LOAF

This recipe is a terrific way to use up your leftover pulp from juicing (see page 232) – saving you money while also reducing food waste. We've used carrot here, but any fruit or veg will do. The loaf is delicious served as is, or topped with nut butter or fruit; see below for some ideas. Store the sliced loaf in an airtight container in the fridge for up to 3 days, or wrap individual slices and freeze for up to 3 months, then thaw in the fridge overnight and enjoy toasted the next day.

MAKES 1 LOAF (10 SLICES)

PREP: 20 MINUTES, PLUS 30 MINUTES RESTING

COOKING: 55 MINUTES

200 g carrots, skins scrubbed, grated

finely grated zest and juice of 1 lemon

2 overripe bananas

¾ cup almond meal

2⅓ cups wholemeal spelt flour

2 teaspoons aluminium-free baking powder

⅓ cup macadamia oil

3–4 tablespoons leftover juice pulp (fruit, vegetable or combo) or water

Preheat the oven to 180°C (160°C fan-forced). Grease and line the base and sides of a 20 cm × 10 cm loaf tin with baking paper.

Add the ingredients, one at a time, to a blender in the order they appear in the ingredients list, adding 3 tablespoons of juice pulp (or water) to begin with. Blend for 30 seconds until smooth and just combined; you don't want to over-blend as this will make the loaf too crumbly. If the mixture looks a bit dry, add another tablespoon of pulp (or water) and blend for 2 seconds only.

Pour the mixture into the prepared tin and level the surface. Bake for 50–55 minutes or until a skewer inserted into the centre comes out clean and the top turns golden.

Remove the loaf from the oven and cool in the tin for 30 minutes before transferring to a wire rack to cool completely. Slice with a serrated knife and serve or store.

6 GREAT TOPPING COMBOS

- *Nut butter, tomato and herbs.*
- *Nut butter and sliced stonefruit.*
- *Avocado, corn kernels and a squeeze of lemon.*
- *Mashed butter beans, shaved fennel and freshly ground black pepper.*
- *Crushed raspberries and a drizzle of tahini.*
- *Coconut yoghurt and banana slices.*

TOFU, SOBA and GREENS MISO SOUP

Although the production of tofu does have a slightly negative effect on our environment, the overall impact is much less severe than most conventional animal agriculture practices. Plus, most of the tofu we eat here is made from Australian soy beans, so food miles are reduced too. This soup is topped with nori seaweed, which is available from supermarkets and Asian grocers in the form of sheets and flakes. Try sprinkling it over salads and soups, or enjoy it on its own as a snack. See pages 121–27 for more about the magic of seaweed.

SERVES 4

PREP: 20 MINUTES

COOKING: 5 MINUTES

⅓ cup shiro (white) miso paste

90 g dried soba noodles, broken in half

1 bunch choy sum, trimmed, torn into thirds

1 bunch thin green asparagus, woody ends discarded, broken into thirds

300 g organic soft tofu, drained, cut into rough 2 cm pieces

4 green onions, thinly sliced into rounds

2 teaspoons sesame seeds, toasted

⅓ cup torn toasted nori

Whisk together the miso paste and 2 litres of water in a large saucepan until well combined. Place the saucepan over medium heat and bring to a simmer.

Add the soba noodles and stir gently with a fork for 1 minute. Add the choy sum and asparagus and allow to simmer gently for 2 minutes or until the green leaves have wilted but the stems are still crunchy.

Divide the tofu and green onion among serving bowls. Ladle over the miso mixture. Sprinkle with the sesame seeds and toasted nori and serve hot.

BAKED BANANA and BLUEBERRY OAT BARS

Packed with antioxidants, bananas and blueberries are terrific for our health. As perennial plants that sequester carbon, they're great for the Earth too. These bars make an excellent on-the-go breakfast or snack. Depending on the season, you can swap the blueberries for grapes, apples or apricots instead.

MAKES 12

PREP: 15 MINUTES, PLUS COOLING TIME

COOKING: 30 MINUTES

- 4 overripe bananas
- 1 teaspoon mixed spice
- 4 large medjool dates, pitted and finely chopped
- 3 cups whole rolled oats
- ⅓ cup pecans, finely chopped
- 125 g blueberries

Preheat the oven to 180°C (160°C fan-forced). Grease and line the base and sides of a 25 cm × 16 cm slice tin with baking paper.

Using a fork, mash the bananas in a large bowl. Add the mixed spice and dates and mix until well combined. Add the oats, pecans and blueberries and stir until well combined.

Press the mixture firmly into the prepared tin, levelling the surface. Bake for 30 minutes or until golden. Remove from the oven and leave to cool completely in the tin before transferring to a board to cut into 12 equal-sized bars. Serve straight away or store in an airtight container in the fridge for up to 4 days.

BROWN RICE PORRIDGE with GRILLED STONEFRUIT

You can use any stonefruit you like in this recipe – apricots, plums, nectarines and peaches all work well. Cultivating rice does release methane into the atmosphere, but new practices are emerging to help lower these emissions. Instead of discarding the cardamom pods after cooking, allow them to cool, then give them a rinse and store them in an airtight container in the freezer. They'll keep for up to 6 months, and you can use them next time you're making Dried spice stock (see page 281).

SERVES 4

PREP: 20 MINUTES

COOKING: 30 MINUTES

- 1½ cups leftover cooked brown rice (basmati is best)
- 1½ cups milk of your choice (try rice, oat, almond or coconut)
- 2 large medjool dates, pitted and finely chopped
- 4 cardamom pods, bruised
- 8 stonefruit, halved and stones removed
- 2 tablespoons pomegranate molasses
- ½ cup flaked almonds

Combine the rice, milk, dates and cardamom in a heavy-based saucepan over medium–low heat. Cook, stirring occasionally, for 25–30 minutes or until the milk has reduced and the mixture is thick and creamy. Remove the pan from the heat and stand, covered, for 3 minutes before removing and setting aside the cardamom pods.

Meanwhile, preheat the grill to high. Line a large baking tray with foil. Place the halved stonefruit, cut-side up, on the prepared tray and grill for 3 minutes. Remove and brush with the pomegranate molasses, then return the stonefruit to the grill for another 3 minutes or until the tops are caramelised. Sprinkle evenly with the almonds and grill for a further 2–3 minutes or until the almonds are golden. (Keep an eye on the fruit as, depending on their ripeness, you may find that some collapse and soften quicker than others.)

Divide the porridge among serving bowls and top with the grilled stonefruit. Serve warm.

CHAI GREEN TEA BREAKFAST FRUIT

All of the fruit in this recipe pulls carbon from the atmosphere while doing very little damage to the soil. The perfect autumn breakfast, this dish can also be made using a slow-cooker – simply halve the amount of liquid and cook for 6–8 hours on high or 10–14 hours on low. Don't forget to keep the whole spices for the Dried spice stock on page 281.

SERVES 4

PREP: 25 MINUTES, PLUS 8 MINUTES STEEPING TIME

COOKING: 30 MINUTES

1 litre boiling water

4 organic, fair-trade green tea teabags

5 cm piece ginger, peeled and thinly sliced

2 cinnamon sticks, broken in half

1 teaspoon black peppercorns

1 teaspoon whole cloves

10 cardamom pods, bruised

1 orange, very thinly sliced into rounds

1 lemon, very thinly sliced into rounds

8 large medjool dates, pitted and quartered lengthways

4 pink-skinned apples, quartered and cored

2 large firm pears, quartered lengthways and seeds removed

1 bunch rhubarb, red part only, cut into 5 cm lengths

Honey-spiced pecans (page 270), to serve

Place the boiling water in a large heatproof jug. Add the teabags and allow to steep for 8 minutes. Remove and discard the teabags.

Pour the brewed tea into a large, deep heavy-based frying pan and add the ginger, cinnamon, peppercorns, cloves, cardamom, orange and lemon. Place over medium heat and allow to come to a simmer. Simmer, untouched, for 10 minutes or until the rinds of the citrus have softened.

Reduce the heat to low. Add the dates, apple, pear and rhubarb. Cover and simmer very gently, turning the fruit just once, for 20 minutes or until very tender and the liquid has reduced by half to a thin syrup. Pick out the whole spices and set aside.

Divide the fruit and syrup among serving bowls. Serve sprinkled with Honey-spiced pecans.

BREKKIE FRIED RICE

Packed with nutritious veg and protein-rich tofu, this brekkie dish is a terrific way to use up leftover rice from last night's dinner.

SERVES 4

PREP: 15 MINUTES

COOKING: 10 MINUTES

2 tablespoons macadamia oil

150 g organic firm tofu, cut into thick matchsticks

1 small red capsicum, cut into matchsticks

200 g peas

3 cups leftover cooked rice (brown basmati or medium-grain brown rice are best)

50 g snow pea sprouts

3 tablespoons tamari or soy sauce

½ teaspoon freshly ground black pepper

2 green onions, thinly sliced diagonally

1 long red chilli, very thinly sliced diagonally (optional)

lemon wedges, to serve

Heat the oil in a large wok over high heat. Add the tofu and stir-fry for 2–3 minutes or until very crisp and golden. Add the capsicum and stir-fry for 30 seconds. Add the peas and stir-fry for 30 seconds.

Add the rice and stir-fry for 2–3 minutes or until hot and starting to crisp. Add the sprouts, tamari or soy sauce and black pepper and stir-fry just until everything is evenly tossed together.

Remove the wok from the heat and toss through the green onion Divide the fried rice among serving plates and sprinkle with the chilli, if desired. Serve hot with lemon wedges alongside.

APPLE and PUMPKIN SCONE

The stars of this scone – apple and pumpkin – sequester carbon while doing very little damage to the soil. You'll need about 300 g of peeled Kent pumpkin to produce the right amount of mash. Simply steam the chopped pumpkin for 12–15 minutes until just tender, then mash until smooth and chill before using. Be careful not to overknead the dough as it will make the scone tough and also prevent it from rising well. Any leftover cooked scone can be wrapped in individual servings and stored in the freezer for 1 month. Thaw in the fridge overnight and toast before serving with savoury or sweet toppings; check pages 234 and 246 for ideas.

SERVES 8–10

PREP: 20 MINUTES, PLUS 3 MINUTES RESTING TIME

COOKING: 25 MINUTES

1 cup mashed cooked and chilled Kent pumpkin (see intro)

1 green apple, coarsely grated

3 tablespoons macadamia oil

2 cups wholemeal spelt flour

2 teaspoons aluminium-free baking powder

1 teaspoon ground cinnamon

½ teaspoon mixed spice

½ teaspoon ground ginger

2 tablespoons milk of your choice (oat, almond or hazelnut)

½ cup pumpkin seeds

Preheat the oven to 180°C (160°C fan-forced). Line a large baking tray with baking paper.

Place the mashed pumpkin, apple and oil in a large bowl and mix until well combined.

In a separate bowl, sift in the flour, baking powder, cinnamon, mixed spice and ginger, tipping any fibre flecks left in the sieve back into the bowl.

Tip half the flour mixture into the mash mixture and, using a large spatula, start to gently fold the mixture together. Once half of the flour mixture has been incorporated, use the spatula in a cutting motion to mix in the rest.

Tip out the mixture onto the prepared tray and begin to bring together with your hands, kneading only very gently until it is just holding shape. Press out to a 2 cm thickness. If your tray is wide enough, press the dough into a 2 cm thick round, and if your tray is not wide enough press the dough out into a 2 cm thick rectangle. The shape doesn't matter here, just the thickness. Using a large knife, score the top of the shaped dough into eight or 10 portions depending on whether it is a circle or rectangle.

Brush the top of the scone with the milk and then sprinkle evenly with pumpkin seeds. Bake for 20–25 minutes or until cooked in the centre and golden on top. Stand on the tray for 3 minutes to rest. Serve warm.

TEFF PANCAKES with BASIL and LIME STRAWBERRIES

Native to Africa, teff is increasingly being grown elsewhere in the world due to its environmental durability. In the US, it is also used as a cover crop to protect soil health. What's more, it's naturally gluten free and packed with protein, iron and fibre. You can find teff grain and flour at health food stores and some larger supermarkets. Give it a go. Serve these pancakes with berries, or try one of the spreads below.

SERVES 4 (MAKES ABOUT 12)

PREP: 15 MINUTES

COOKING: 35 MINUTES

1 cup teff flour

2 teaspoons aluminium-free baking powder

1 large overripe banana

2 large free-range eggs or ½ cup sugar-free apple puree

1 cup milk of your choice (hazelnut, almond and coconut all work well)

macadamia oil, for cooking

coconut yoghurt, for serving (optional)

BASIL AND LIME STRAWBERRIES

1 sprig basil, leaves picked, stems trimmed and thinly sliced

250 g strawberries, hulled and thinly sliced

finely grated zest and juice of 1 lime

To make the basil and lime strawberries, place all the ingredients in a bowl and stir to combine. Set aside at room temperature to macerate, stirring occasionally.

Place the teff, baking powder, banana, eggs or apple puree and milk in a blender and blend for 30–40 seconds until smooth.

Heat a large cast-iron frying pan over medium heat. Once hot, brush the base with oil and pour in enough of the teff batter to form two 8 cm rounds. Cook for 2–3 minutes or until the pancakes are firm to the touch around the edges and the centres have set (they need to be firm to be able to flip without breaking). Carefully flip and continue to cook for 1–2 minutes or until golden. Transfer to a plate and cover with a clean tea towel to keep warm. Repeat the process to make about 12 pancakes in total.

Divide the pancakes among serving plates, top with coconut yoghurt (if using) and spoon over the basil and lime strawberries. Serve warm.

4 ANYTIME SPREADS

- **Apple almond butter:** *Soak 1 cup blanched almonds in water overnight, then drain and rinse. Blend with 1 small grated apple and ½ teaspoon ground cinnamon until smooth.*
- **Strawberry chia jam:** *Blend 250 g strawberries, 1 large pitted medjool date and 3 teaspoons chia seeds until smooth. Pour into a jar and stand for 5 minutes to set.*
- **Lemon macadamia spread:** *Soak 1 cup macadamias in water overnight, then drain and rinse. Blend with the zest of 2 lemons and 2 tablespoons lemon juice until smooth.*
- **Choc coconut spread:** *Blend 400 ml coconut milk, 2 large pitted medjool dates and ⅓ cup raw cacao powder until smooth. Pour into two jars and chill for 2–3 hours or until set to a spreadable consistency.*

LUN

CH

WALNUT and CARROT BUCKWHEAT SALAD

Buckwheat is versatile, nutritious and delicious. It is also a great cover crop, plus it can help suppress weeds – resulting in fewer chemicals being used on the land.

SERVES 4

PREP: 30 MINUTES, PLUS 5 MINUTES RESTING TIME

COOKING: 40 MINUTES

- 3 tablespoons avocado oil
- 1 garlic clove, crushed
- 1 teaspoon ground cinnamon
- 1 red onion, very thinly sliced
- 1 cup raw buckwheat
- 1½ cups stock of your choice (pages 280–81)
- sea salt and freshly ground black pepper
- 1 bunch rhubarb, red part only, trimmed and halved crossways
- 2 bunches baby carrots, tops trimmed, skins scrubbed, halved lengthways
- ½ cup walnut pieces
- 1 pomegranate, halved, seeds tapped out
- 2 cups mixed leaves
- lemon wedges, to serve

Heat the oil in a saucepan over medium heat. Add the garlic, cinnamon and onion and cook, stirring occasionally, for 8–10 minutes or until the onion is golden and slightly crisp.

Using a fork, stir the buckwheat and stock into the onion mixture in the pan and continue stirring until it comes to a simmer. Immediately reduce the heat to the lowest possible setting and cook, covered, for 25–30 minutes or until the buckwheat is cooked and the stock has been completely absorbed.

Remove the pan from the heat and stand, covered, for 5 minutes before fluffing and separating the grains with a fork. Season well. Transfer the buckwheat mixture to a large bowl.

Meanwhile, preheat the oven to 200°C (180°C fan-forced). Place the rhubarb and carrot on a large baking tray, season well and bake for 15 minutes. Scatter over the walnuts and bake for a further 5–10 minutes or until the carrot is just tender and the walnuts are golden; the rhubarb will be very soft but still hold its shape.

Add the pomegranate seeds, mixed leaves and the roasted rhubarb mixture to the buckwheat mixture in the bowl and lightly toss together. Divide evenly among serving plates. Serve warm with lemon wedges.

TEMPEH SATAY with LIME ZOODLES

Tempeh, made from fermented soybeans, is a brilliant plant-based protein. Like tofu, the stuff we eat in Australia is often made from locally grown soy beans, which cuts down on food miles too. Freshly churned peanut butter is the way to go with this satay. It's available from bulk-food stores or you can churn your own by blitzing 450 g roasted peanuts in a food processor for 5 minutes (just make sure you keep blitzing for this length of time, as you need the natural oils from the peanuts to be released).

SERVES 4

PREP: 30 MINUTES

COOKING: 10 MINUTES

4 zucchini, spiralised

2 carrots, spiralised

300 g daikon, peeled and spiralised

finely grated zest and juice of 2 limes

1 cup small coriander sprigs

½ cup small mint sprigs

sea salt and freshly ground black pepper

2 tablespoons macadamia oil

300 g tempeh, thinly sliced

1 long red chilli, thinly sliced

4 green onions, cut into 2 cm lengths

⅔ cup peanut butter

400 ml can coconut cream

Place the zucchini, carrot, daikon, lime zest and juice, coriander and mint in a large bowl and gently toss to combine. Season well and set aside.

Heat the oil in a large wok over high heat. Add the tempeh and stir-fry for 3–5 minutes or until heated through and very crisp. Using tongs, transfer the tempeh to a plate.

Reduce the heat to medium and add the chilli and green onion. Stir-fry for 20 seconds or until fragrant, then add the peanut butter and coconut cream and gently stir until well combined and heated through. Remove the wok from the heat and add the tempeh to the satay mixture, stirring well to combine.

Pile the zoodle salad over the tempeh satay and serve warm.

FLAVOUR BOOSTERS

A spoonful or two of these will give a lift to your leftovers. Arame is a type of seaweed, which we now know is magnificent in many ways, while the tapenade contains olives, one of the highest sequesters of carbon. So eating more of these would be great for the planet. Go on.

DILL PICKLED CUCUMBER

Combine 3 very thinly sliced Lebanese cucumbers, 2 teaspoons toasted brown mustard seeds, finely chopped stems and leaves from 1 small sprig of dill and 3 tablespoons apple cider vinegar and season well. Leave at room temperature, turning occasionally, for 10–15 minutes or until lightly softened. Store in an airtight container in the fridge for up to 3 days (where they will continue to soften).

PICKLED ARAME

Soak ½ cup organic arame in warm water for 5 minutes to soften, then drain well. Combine with 1 tablespoon tamari, 1 cm piece ginger, peeled and grated, 2 thinly sliced green onions and 2 teaspoons toasted sesame seeds. Leave at room temperature for 5 minutes. Store in an airtight container in the fridge for up to 2 days.

EGGPLANT AND FIG CHUTNEY

In a heavy-based saucepan over medium heat cook 2 tablespoons macadamia oil, 1 finely chopped red onion, 4 thinly sliced baby eggplants, 1 chopped long red chilli, leaves from 1 sprig curry leaves and 2 teaspoons garam masala for 8–10 minutes or until very soft and light golden. Stir in 1 chopped large, overripe tomato and 3 tablespoons red wine vinegar and simmer for 5 minutes. Stir in 4 chopped figs, then remove the pan from the heat and stand, covered, stirring occasionally, until cooled to room temperature. Store in an airtight container in the fridge for up to 4 days.

PISTOU

Using a mortar and pestle, pound 1 clove garlic and 2 teaspoons sea salt until a paste forms. Add the leaves from 2 sprigs of basil and pound together, grinding the leaves against the side of the mortar until a paste-like puree forms. Slowly incorporate ⅓ cup extra virgin olive oil into the paste, a tablespoon at a time, until a creamy sauce forms. Season with pepper. Store in an airtight container in the fridge for up to 2 days.

BEETROOT TAPENADE

Combine 1 finely chopped roasted and peeled beetroot, ½ cup finely chopped Sicilian olives, 1 tablespoon finely chopped rinsed salted capers, 1 tablespoon finely chopped rosemary, 1 tablespoon finely chopped oregano and 3 tablespoons extra virgin olive oil, then season well. Store in an airtight container in the fridge for up to 4 days.

CAULIFLOWER AND MACADAMIA 'HUMMUS'

Steam 2 cups cauliflower florets and 1 cup macadamias for 15–18 minutes or until tender. Cool slightly. Transfer the macadamias to a blender and process until finely chopped, then add the cauliflower, 1 tablespoon tahini and the finely grated zest and juice of 1 small lemon. Blend until smooth and season well. Transfer to a bowl and gently stir through 1 tablespoon avocado oil and a large pinch of sweet paprika. Store in an airtight container in the fridge for up to 2 days.

ITALIAN BEAN SOUP

Beans are a great substitute for conventionally produced meat, as they have a much lower environmental impact. High in protein and budget-friendly, they are also perfect for batch-cooking and freezing, saving both time and energy.

SERVES 4–6

PREP: 25 MINUTES

COOKING: 30 MINUTES

3 tablespoons extra virgin olive oil, plus extra to serve

1 large onion, finely chopped

2 garlic cloves, crushed

3 tablespoons tomato paste

2 carrots, scrubbed and finely chopped

2 celery stalks, finely chopped

1 zucchini, finely chopped

1 small bulb fennel, finely chopped

4 sprigs basil, leaves picked and stems finely chopped

1.5 litres stock of your choice (pages 280–81)

400 g can cannellini beans, drained and rinsed

400 g can borlotti beans, drained and rinsed

sea salt and freshly ground black pepper

Heat the oil in a large saucepan over medium heat. Add the onion and cook, stirring occasionally, for 5 minutes or until very soft and light golden. Add the garlic and tomato paste and cook, stirring constantly, for 3 minutes or until the tomato paste is rich and thick and starting to separate in the pan.

Add all the remaining ingredients to the pan and cook, stirring, until the mixture comes to a simmer. Partially cover the pan and continue to simmer, stirring occasionally, for 18–20 minutes or until the vegetables are very tender. Season well.

Divide the soup among serving bowls. Finish with a drizzle of olive oil and serve.

ROAST BEETROOT and QUINOA SALAD with TURMERIC DRESSING

Native to South America, quinoa is now grown closer to home in Tasmania and Western Australia. As a complete protein containing all nine essential amino acids, it's a terrific alternative to animal products. To toast the seeds for the turmeric dressing, gently cook them in a heavy-based pan over very low heat for 1–2 minutes or until you begin to smell their aroma and they turn very light golden – just be careful not to take them too far as they can become burnt and bitter quite quickly.

SERVES 4

PREP: 35 MINUTES

COOKING: 40 MINUTES

2 bunches baby beetroot, scrubbed, halved lengthways

2 small bulbs fennel, sliced

1 red onion, quartered

3 tablespoons extra virgin olive oil

1 sprig rosemary, leaves stripped

sea salt and freshly ground black pepper

½ cup brazil nuts, chopped

½ cup quinoa, rinsed and drained well

¾ cup stock of your choice (pages 280–81)

TURMERIC DRESSING

3 tablespoons apple cider vinegar

2 tablespoons extra virgin olive oil

2 teaspoons freshly grated turmeric or ½ teaspoon ground turmeric

½ cup coconut yoghurt

2 teaspoons caraway seeds, toasted

1 teaspoon brown mustard seeds, toasted

Make the turmeric dressing first by whisking all the ingredients until smooth and well combined. Add a little water to loosen if necessary, so that the dressing is of a thick pouring consistency. Season well.

Preheat the oven to 200°C (180°C fan-forced).

Place the beetroot, fennel, onion, oil and rosemary in a large bowl. Season well. Toss together until well combined and coated. Spread the mixture evenly over a large baking tray, setting the bowl aside. Bake for 35 minutes, then add the brazil nuts to the tray and bake for a further 5 minutes or until the beetroot is tender and everything is golden.

Meanwhile, place the quinoa and stock in a small saucepan over medium heat. Using a fork, stir together until the mixture comes to a simmer, then immediately reduce the heat to the lowest setting possible and cook, covered, for 15–18 minutes or until the quinoa is cooked and the stock has been completely absorbed.

Remove the pan from the heat and stand, covered, for 5 minutes before fluffing and separating the grains with a fork. Add the quinoa to the bowl that the vegetables were tossed in, then add the cooked vegetables straight from the oven, tossing gently until just combined.

Divide the salad among plates and serve drizzled with the turmeric dressing.

CAULIFLOWER and POTATO SOUP with DUKKAH ALMONDS

Ideal for lunch or a light dinner on cooler days, this soup is really easy to prepare. The dukkah almonds are an optional but delicious addition, and are excellent as a standalone snack too.

SERVES 4

PREP: 25 MINUTES, PLUS 5 MINUTES STANDING TIME

COOKING: 25 MINUTES

- 2 tablespoons macadamia oil
- 1 large onion, chopped
- 1 (550 g) small cauliflower, stems chopped and florets broken into pieces
- 2 potatoes, scrubbed and chopped
- 1.5 litres stock of your choice (pages 280–81)
- sea salt and freshly ground black pepper
- ½ teaspoon saffron threads (optional)

DUKKAH ALMONDS

- 1 cup almonds
- 2 tablespoons macadamia oil
- 2 tablespoons dukkah
- 1 teaspoon sea salt flakes

To make the dukkah almonds, preheat the oven to 180°C (160°C fan-forced). Line a baking tray with baking paper. Toss all the ingredients together in a bowl until well coated. Spread evenly over the prepared tray and bake for 12–15 minutes or until golden. Remove from the oven and cool on the tray.

Meanwhile, heat the oil in a large saucepan over medium heat. Add the onion and cook, stirring occasionally, for 5 minutes or until softened. Add the cauliflower, potato and stock and stir until the mixture comes to a simmer. Continue to simmer, partially covered, for 15–18 minutes or until the potato is soft. Remove the pan from the heat and stand for 5 minutes.

Using a hand-held stick blender, blend the potato mixture until completely smooth. Season well.

Divide the soup among serving bowls and top with the saffron, if you like. Serve hot scattered with the dukkah almonds.

KALE 'FRITTATA' with NUT BASE

This dairy-free 'frittata' has a significantly lower environmental impact than a traditional version made with eggs and cheese. If you've not come across nutritional yeast flakes before, why not give them a try? Buy a small amount from a bulk-food shop and see what you think. Otherwise, you can simply leave them out – this dish will still be delicious.

SERVES 8

PREP: 25 MINUTES, PLUS 25 MINUTES STANDING TIME

COOKING: 45 MINUTES

2 tablespoons extra virgin olive oil

1 garlic clove, crushed

4 green onions, thinly sliced

4 large kale leaves, stems finely chopped and leaves torn

1½ cups roughly chopped mixed nuts (macadamias, almonds, brazil nuts)

500 g organic firm tofu, drained well

½ teaspoon freshly grated turmeric or ¼ teaspoon ground turmeric

2 tablespoons nutritional yeast flakes (optional)

sea salt and freshly ground black pepper

Heat the oil in a saucepan over medium heat. Add the garlic, green onion and kale and cook, stirring, for 3 minutes or until cooked but not golden. Transfer the mixture to a bowl and allow to cool completely.

Meanwhile, preheat the oven to 180°C (160°C fan-forced). Liberally grease the base and side of a glass pie plate, about 20 cm across and 4 cm deep.

Spread the nuts evenly over the base of the prepared pie plate and bake for 8 minutes or until just a very light golden colour. Remove from the oven and stand for 20 minutes to cool.

Break the tofu into large pieces and place in the bowl of a food processor, along with the turmeric and nutritional yeast, if using. Season well. Process until smooth. Add the cooled kale mixture and, using the pulse button, process in bursts for 20–30 seconds or until well combined.

Carefully spread the tofu mixture over the nut base in the pie plate and level the surface. Bake for 30–35 minutes or until set firm and golden. Stand for 3 minutes before slicing. Serve warm.

CHARGRILLED CABBAGE and PUMPKIN FREEKEH SALAD with HERB STEM DRESSING

This vibrant, veggie-packed salad is accompanied by a clever dressing that puts to use all those herb stems that would usually end up in landfill, releasing damaging methane into the atmosphere.

SERVES 4

PREP: 30 MINUTES

COOKING: 30 MINUTES

¾ cup freekeh

½ small red cabbage, cut into 8 wedges

½ small Kent pumpkin, cut into 8 wedges, skin left on, seeds intact

2 zucchini, very thinly sliced into rounds

½ cup pistachio kernels, toasted and roughly chopped

sea salt and freshly ground black pepper

HERB STEM DRESSING

3 tablespoons white wine vinegar

2 tablespoons avocado oil

1 teaspoon Dijon mustard

1 small red onion, finely chopped

3 dill stems, finely chopped

2 basil stems, finely chopped

4 flat-leaf parsley stems, finely chopped

Make the herb stem dressing by combining all the ingredients in a small bowl. Season well. Set aside at room temperature until ready to serve, stirring occasionally.

Place the freekeh in a saucepan and cover with cold water. Place the pan over medium–high heat and bring to a rapid simmer, using a fork to stir occasionally. Reduce the heat to medium and simmer for 30 minutes, untouched, or until the freekeh is cooked. Drain well, then immediately rinse under cold running water to cool. Drain well again before transferring to a large bowl. Set aside.

Meanwhile, preheat a large chargrill pan over medium–low heat. Chargrill the cabbage and pumpkin, in two batches, for 8 minutes or until cooked and golden.

Transfer the cooked cabbage and pumpkin to the bowl with the freekeh and add the zucchini, pistachios and the herb stem dressing. Toss really well to combine. Season well.

Divide the salad among serving plates. Serve while the vegetables are still warm.

SPLIT PEA and VEGETABLE SOUP

Simple food at its best. Soaking the split peas overnight helps them to soften, reducing their cooking time. If you forget, don't worry: just increase the stock to 2 litres, then cover and simmer the soup mixture for 2–2½ hours instead. To freeze the soup, pour cooled single serves into airtight containers and keep for up to 6 months. Thaw in the fridge overnight and reheat gently on the stovetop, adding a little extra stock or water to loosen, if needed.

SERVES 4–6

PREP: 15 MINUTES, PLUS OVERNIGHT SOAKING TIME

COOKING: 45 MINUTES

500 g split green peas

2 tablespoons extra virgin olive oil

2 leeks, white part sliced

2 garlic cloves, crushed

2 carrots, chopped

2 zucchini, chopped

2 celery stalks, chopped

1.5 litres stock of your choice (pages 280–81)

1 cup flat-leaf parsley leaves

sea salt and freshly ground black pepper

Soak the split peas overnight in a very large bowl of water, covered and at room temperature. Rinse and drain very well.

Heat the oil in a large saucepan over medium heat. Add the leek and garlic and cook, stirring occasionally, for 5 minutes or until the leek is very soft. Add the carrot, zucchini, celery, stock and drained peas. Cook, stirring constantly, until the mixture comes to a simmer. Reduce the heat to medium–low and simmer gently, partially covered, for 40 minutes or until the peas have softened completely, stirring occasionally to make sure the peas don't stick to the base of the pan.

Remove the pan from the heat and stir through the parsley. Season well.

Ladle the soup into bowls and serve hot.

MEXICAN ZUCCHINI and RED KIDNEY BEAN FRITTERS

Full of protein, vitamins and minerals, red kidney beans are also often free of pesticides due to the way they are harvested. A great meat-free midweek option.

SERVES 4

PREP: 35 MINUTES

COOKING: 25 MINUTES

- 2 tablespoons chia seeds
- 400 g can red kidney beans, drained and rinsed
- 1 small zucchini, grated
- sea salt and freshly ground black pepper
- 1 cup wholemeal spelt flour
- 2 teaspoons aluminium-free baking powder
- 2 teaspoons ground cumin
- 1 teaspoon smoked paprika
- ⅔ cup milk of your choice
- macadamia oil, for cooking

AVOCADO SALAD

- finely grated zest and juice of 2 limes
- 1 long red chilli, finely chopped
- 2 tablespoons avocado oil
- 2 green onions, thinly sliced into rounds
- 1 corn cob, husk and silk discarded and kernels removed
- 4 sprigs coriander, washed well, roots and stems finely chopped, leaves picked
- 2 avocados, halved, stones removed and skin peeled

Place the chia seeds and ½ cup of water in a large bowl. Using a fork, whisk together, then set aside for 1 minute. Whisk again and then add the kidney beans and mash thoroughly. Add the zucchini and mix until well combined. Season well. Add the flour, baking powder, cumin and paprika and mix well to combine. Slowly stir through the milk until the batter is well combined and smooth.

Heat a large cast-iron frying pan over medium–high heat and add some oil to coat the base. Using a ¼ cup measure, drop four measures of the batter into the hot pan and shape into roughly 7 cm rounds. Cook for 3–4 minutes or until set underneath and starting to turn golden around the edges, and small bubbles appear and burst on the surface. Carefully flip and cook for a further 3–4 minutes, or until cooked and golden. Transfer to a plate and cover with a clean tea towel to keep warm. Repeat twice more with the remaining oil and batter to make 12 fritters in all. Keep warm.

To make the avocado salad, gently combine the lime zest and juice, chilli, oil, green onion, corn and coriander in a bowl. Season well. Divide the avocado halves among serving plates and spoon over the corn mixture.

Serve the fritters warm, alongside the avocado salad.

MARINATED BUTTER BEAN and RAS EL HANOUT EGGPLANT SALAD

Pulses such as beans are an environmentally friendly (and budget-friendly) alternative to conventional animal protein. Ras el hanout is a spice blend commonly used in Moroccan cooking. You'll find it in larger supermarkets and some bulk-food stores. Try sprinkling it over your favourite veg before baking, chargrilling or barbecuing.

SERVES 4

PREP: 30 MINUTES

COOKING: 30 MINUTES

- ⅓ cup extra virgin olive oil
- 3 teaspoons ras el hanout
- 2 eggplants, cut into rough 4 cm pieces
- sea salt and freshly ground black pepper
- 1 tablespoon wholegrain mustard
- 3 tablespoons apple cider vinegar
- 400 g can butter beans, drained and rinsed
- 2 sprigs flat-leaf parsley, stems and leaves finely chopped
- 1 bunch red radishes, trimmed and very thinly sliced
- 2 yellow squash, very thinly sliced into rounds
- 2 oranges, zest finely grated, then peel removed and flesh sliced into thin rounds

Preheat the oven to 200°C (180°C fan-forced).

Place 2 tablespoons of the oil, the ras el hanout and eggplant in a bowl. Season well. Toss to combine and coat well, then spread evenly over a large baking tray. Bake for 25–30 minutes or until the eggplant is very tender and caramelised.

Meanwhile, place the mustard, vinegar and remaining oil in a large bowl and, using a fork, whisk until well combined. Add the butter beans and toss to coat in the mixture. Set aside to marinate at room temperature while the eggplant cooks.

Add the hot, cooked eggplant to the butter bean mixture and toss until well combined. Add the parsley, radish, squash and orange zest and rounds, and gently toss together. Season well.

Serve warm.

NUT-TASTIC! SNACKS

Coconut and cacao are excellent carbon-sequesterers. Whenever possible, go for locally grown nuts over imports. It's also worth noting that it can take a lot of water to produce almonds, so they're best used sparingly. Honey is the greenest choice when it comes to sweeteners (sugar is the most environmentally destructive), plus our bees are in desperate need of help. Always choose local, sustainable brands when you can and enjoy in moderation (I couldn't help myself).

HONEY-SPICED PECANS

Toss together 1 cup pecans, 1 teaspoon mixed spice, 1 tablespoon honey and 2 teaspoons macadamia oil. Spread in a single layer over a lined baking tray and cook in a 180°C (160°C fan-forced) oven for 10–12 minutes or until golden. Cool on the tray. Store in an airtight container for up to 1 week. Enjoy as a snack or use in the Chai green tea breakfast fruit on page 240.

PEANUT BUTTER COOKIE DOUGH BALLS

Combine 6 large pitted medjool dates with ⅓ cup boiling water and stand, covered with a clean tea towel, for 5 minutes or until the dates soften and the water cools slightly. Transfer the mixture to a food processor and add ½ cup peanut butter and 1 cup oat flour. Blend until smooth, then roll into 20 large balls. Roll the balls in chia seeds, shredded coconut or cacao nibs. Place on a lined baking tray and freeze for 20 minutes or until slightly firm.

WALNUT FREEZER COOKIES

Blend 1½ cups walnuts in a food processor until finely chopped. Add ¾ cup coconut flakes, ½ teaspoon mixed spice and 1 tablespoon honey. Process for 1–2 minutes or until the mixture comes together into a ball. Shape into cookies and transfer to a lined baking tray. Freeze for 20 minutes or until firm. You can store these in an airtight container in the freezer for up to 2 months.

MACADAMIA CREAM

Soak 1 cup macadamias in water overnight to soften, then drain and rinse well. Blend with 1 teaspoon ground ginger, 2 teaspoons honey and 1–2 tablespoons milk of your choice (almond or coconut work well) until completely smooth.

CHOC-CHILLI BRAZIL NUTS

Dip lightly toasted brazil nuts in melted 70 per cent dark chocolate to coat two-thirds, then transfer to a lined baking tray. Sprinkle with finely chopped red chilli and leave to set firm, in the fridge if required.

ALMOND-STUFFED DATES

Stuff large pitted medjool dates with a little tahini and roasted almonds – that's it!

DIN

NER

SPRING LENTILS

Lentils are a win for our environment. They 'fix' nitrogen in our soil (so don't require nitrogen fertilisers), plus they are a very water-efficient source of protein, requiring around 350 litres of water to grow 1 kg, compared with around 15,000 litres for 1 kg of conventionally farmed beef.

SERVES 4

PREP: 25 MINUTES, PLUS 5 MINUTES RESTING TIME

COOKING: 45 MINUTES

3 tablespoons extra virgin olive oil

2 large garlic cloves, crushed

2 small fennel bulbs, finely chopped

1 large carrot, scrubbed and finely chopped

1 large leek, white part thinly sliced

1 cup puy (French-style) lentils

4 fresh bay leaves

8 sprigs thyme

1 litre stock of your choice (pages 280–81)

4 large kale leaves, white stems removed, leaves thickly shredded

1 bunch thin green asparagus, woody ends discarded, broken into thirds

¾ cup peas

sea salt and freshly ground black pepper

lemon wedges, to serve

Heat the oil in a large, deep heavy-based frying pan over medium–high heat. Add the garlic, fennel, carrot and leek and cook, stirring occasionally, for 5 minutes or until softened and very light golden. Add the lentils and cook, stirring constantly, for 1 minute or until they are well coated in the oil mixture.

Reduce the heat to medium–low and stir in the bay leaves, thyme and stock. Partially cover and simmer gently, stirring occasionally, for 30 minutes or until just tender. Add the kale and stir until it just wilts, then add the asparagus and peas and stir well. Immediately remove the pan from the heat and set aside, covered, for 5 minutes. There will be enough heat in the pan to lightly soften the asparagus and peas. Season well.

You can take the pan straight to the table or divide the lentil mixture among serving bowls. Serve hot with lemon wedges.

STOVE-TOP BEAN CASSOULET with NUT CRUMB

The combination of beans and nuts here makes this a superb 'emission-friendly' meal – and it's even better if the ingredients are locally sourced.

SERVES 4

PREP: 15 MINUTES

COOKING: 55 MINUTES

3 tablespoons extra virgin olive oil

2 onions, chopped

3 garlic cloves, chopped

4 sprigs thyme

2 sprigs rosemary

400 g can whole peeled tomatoes

2 large zucchini, chopped into bite-sized pieces

2 × 400 g cans cannellini beans, drained and rinsed

2 cups stock of your choice (pages 280–81)

lemon wedges, to serve

NUT CRUMB

½ cup walnuts, toasted and finely chopped

½ cup almonds, toasted and finely chopped

½ cup macadamias, toasted and finely chopped

1 sprig flat-leaf parsley, stems and leaves finely chopped

sea salt and freshly ground black pepper

Make the nut crumb by placing all the ingredients in a bowl and stirring until well combined. Season to taste. Set aside.

Heat the oil in a large, deep heavy-based frying pan over medium heat. Add the onion and cook, stirring occasionally, for 15 minutes or until the onion has collapsed and is very golden and starting to catch on the bottom of the pan. (This is important, as this long cooking time adds a crucial depth of flavour and deep colour to the cassoulet.)

Add the garlic, thyme and rosemary and cook, stirring, for 1 minute or until fragrant. Add the tomatoes and zucchini and cook, breaking up the tomato pieces roughly with the back of your spoon, for 30 seconds or until well coated in the onion mixture.

Reduce the heat to low and add the beans and stock. Partially cover and simmer, stirring occasionally, for 30–35 minutes or until the sauce has reduced by three-quarters and the mixture is rich and thick.

Take the pan straight to the table and sprinkle the nut crumb evenly over the top. Serve with lemon wedges.

BARBECUED VEGETABLES and TEMPEH with TAHINI DRESSING

In the warmer months, when zucchini, eggplant and asparagus are in season (and at their best and cheapest), chuck them on the barbecue and let them shine in this flavour-packed salad. Tempeh provides the protein here, along with the creamy tahini dressing. Try serving this with the pickled arame on page 254 – a terrific way of incorporating more beautiful seaweed into your life!

SERVES 4

PREP: 25 MINUTES

COOKING: 10 MINUTES

2 tablespoons avocado oil

2 teaspoons smoked paprika

300 g tempeh, cut into 1 cm thick slices

1 red onion, cut into wedges

6 baby eggplants, halved lengthways

4 small zucchini, halved lengthways

sea salt and freshly ground black pepper

2 bunches asparagus, woody ends discarded

4 large kale leaves, white stems removed and leaves torn

Pickled arame, to serve (page 254)

TAHINI DRESSING

1 tablespoon tahini

2 tablespoons avocado oil

½ teaspoon sweet paprika

finely grated zest and juice of 1 large lemon

2–3 tablespoons warm water

Preheat a barbecue chargrill plate to medium–high.

Place the oil, paprika, tempeh, onion, eggplant and zucchini in a large bowl and season to taste. Toss well to combine. Transfer to the hot chargrill and cook, turning occasionally, for 3–5 minutes or until cooked and golden. Return to the bowl and toss gently.

Add the asparagus and kale to the chargrill and cook, turning only once, for 2 minutes or until the asparagus is just tender and charred and the kale is crisp. Add to the bowl with the vegetables and toss gently until just combined. Divide evenly among serving plates or arrange on a platter to share.

To make the tahini dressing, use a fork to whisk all the ingredients until smooth and well combined, adding more warm water if needed to make it a pouring consistency. Season to taste.

Drizzle the tahini dressing over the salad. Serve warm with the pickled arame alongside.

STOCK FOUR WAYS

These homemade stocks will produce about 2 litres each. The colour and flavour will be more subtle than commercially made stock, so remember to season them well before using.

After straining, you can either use the stock immediately or cool it to room temperature and divide into 1 cup portions in airtight containers. They'll keep in the freezer for up to 6 months.

WINTER VEGETABLE STOCK

Combine 6 cups winter vegetable peelings (pumpkin, cabbage, fennel etc), 1 chopped onion and 1 chopped carrot in a large saucepan and cover with 3 litres water. Boil for 45 minutes or until reduced by a third. Stand, covered, for 20 minutes before straining.

SUMMER VEGETABLE STOCK

Combine 6 cups summer vegetable peelings (asparagus, beans, zucchini etc), 1 chopped onion and 1 chopped carrot in a large saucepan and cover with 3 litres water. Boil for 45 minutes or until reduced by a third. Stand, covered, for 20 minutes before straining.

HERB AND SPICE STOCK

Combine 4 cups chopped herb stalks, 1 head garlic, 3 cm piece ginger, 3 cm piece turmeric and 8 chopped green onions in a large saucepan and cover with 3 litres water. Boil for 45 minutes or until reduced by a third. Stand, covered, for 20 minutes before straining.

DRIED SPICE STOCK

Combine 1 cup mixed whole dried spices (cinnamon sticks, cardamom pods, cumin seeds, coriander seeds, star anise), 2 chopped onions, 1 chopped carrot and 2 chopped celery stalks in a large saucepan and cover with 3 litres water. Boil for 45 minutes or until reduced by a third. Stand, covered, for 20 minutes before straining.

STEMS and LEAVES RISOTTO

A simple, nutritious meal that also helps to reduce your food waste. Celery heart is the tender inner section of a bunch of celery, and has a sweeter, more delicate flavour. The pale inner leaves of the cauliflower and broccoli are those that you find still snugly attached after you have removed the larger outer leaves.

SERVES 4

PREP: 25 MINUTES, PLUS 5 MINUTES RESTING TIME

COOKING: 35 MINUTES

3 tablespoons avocado oil

2 garlic cloves, crushed

1 leek, white part sliced

1 celery heart, stalks sliced, inner leaves separated

1¼ cups arborio rice

1.5 litres stock of your choice, heated (pages 280–81)

150 g cauliflower stems, finely chopped

150 g broccoli stems, finely chopped

1 sprig basil, leaves picked, stems chopped

1 sprig flat-leaf parsley, leaves picked, stems chopped

1 sprig dill, leaves picked, stems chopped

finely grated zest and juice of 1 lemon

4 inner cauliflower leaves

8 inner broccoli leaves

½ cup hazelnuts, toasted, skins removed, chopped

sea salt and freshly ground black pepper

Heat 2 tablespoons of the oil in a heavy-based saucepan over medium heat. Add the garlic, leek and celery stalks and cook, stirring occasionally, for 3 minutes or until almost softened. Add the rice and cook, stirring constantly, until the grains are well coated in the oil mixture.

Reduce the heat to low, stir in one ladleful of the heated stock and cook, stirring slowly, for 2–3 minutes or until the stock is completely absorbed before adding the next ladleful. Continue cooking while adding ladles of stock. When you have used half the stock, stir in the cauliflower stem, broccoli stem, basil stem, parsley stem and dill stem. Continue to cook, adding the stock and stirring slowly between each addition, until the rice is just tender. This will take about 18–25 minutes, depending on your rice, and you may find that you don't need to use all the stock. Remove the pan from the heat and set aside, covered, for 5 minutes.

Meanwhile, place the lemon zest, cauliflower leaves, broccoli leaves, hazelnuts, basil leaves, parsley leaves, dill leaves, celery heart leaves and the remaining oil together in a bowl and toss well to combine. Season well.

Stir the lemon juice through the risotto mixture and season well. Divide evenly among bowls or plates and top with the leaves mixture. Serve immediately.

JERUSALEM ARTICHOKE and MUSHROOM BOURGUIGNON

Jerusalem artichokes are an excellent source of nectar for pollinating insects such as bees and butterflies. If you've not tried them before, this wonderfully hearty and comforting meal is a great place to start.

SERVES 4

PREP: 25 MINUTES

COOKING: 40 MINUTES

- 2 tablespoons extra virgin olive oil
- 4 field mushrooms, trimmed
- 8 portobello mushrooms, trimmed
- sea salt and freshly ground black pepper
- 1 carrot, scrubbed and chopped
- 1 large onion, chopped
- 450 g Jerusalem artichokes, peeled and halved lengthways
- 1 bunch bulb spring onions, tops thinly sliced, bulbs trimmed and left whole
- 2 garlic cloves, crushed
- 2 tablespoons tomato paste
- 4 fresh bay leaves
- 1 cup red wine
- 1 cup stock of your choice (pages 280–81)
- steamed greens (such as green beans and zucchini), to serve

Heat half the oil in a large, deep heavy-based frying pan over high heat and brown the field and portobello mushrooms for 1 minute, then remove to a bowl, season well and set aside.

Reduce the heat to medium and add the remaining oil. Add the carrot, onion and artichokes and cook, stirring occasionally, for 5 minutes or until the onion has softened and the vegetables are golden. Add the spring onion, garlic and tomato paste and cook, stirring, for 3 minutes or until thick and rich.

Reduce the heat to low, add the bay leaves, red wine and stock, and stir until well combined. Simmer gently, stirring occasionally, for 25 minutes. Return the mushrooms to the pan, along with any resting juices in the bowl, and cook for 5 minutes or until the vegetables are tender and the sauce reduced by three-quarters. Remove the pan from the heat and season well.

Take the pan straight to the table and serve with the steamed vegetables alongside.

LEFTOVER VEGGIE PASTA SAUCE

Another brilliant way to save money and reduce your food waste.

SERVES 4

PREP: 10 MINUTES

COOKING: 10 MINUTES

- 2 tablespoons extra virgin olive oil, plus extra to serve
- 2 garlic cloves, sliced
- 1 sprig basil, stems and leaves finely chopped
- 400 g can whole peeled tomatoes
- 2 cups leftover cooked vegetables (steamed or roasted pumpkin, zucchini, onion, cauliflower)
- sea salt and freshly ground black pepper
- cooked spelt spiral pasta, to serve

Heat the oil in a cast-iron frying pan over medium heat. Add the garlic, basil and tomatoes and cook, stirring and breaking up the tomato using the back of your spoon, for 5 minutes or until reduced by a quarter.

Add the leftover vegetables and cook, stirring occasionally, for 5 minutes or until heated through and the sauce has reduced by half. Season well.

Divide the pasta among serving bowls and spoon over the pasta sauce. Finish with an extra drizzle of oil and serve.

MOROCCAN ROASTED CAULI and APRICOTS with ZUCCHINI COUSCOUS

In this grain-free zucchini 'couscous' recipe, we use whole herbs rather than just picking the leaves and discarding the stems, which then end up in landfill. Go to pages 262 and 282 to check out more fantastic herb-stem ideas.

SERVES 4

PREP: 35 MINUTES

COOKING: 45 MINUTES

ROASTED CAULI AND APRICOTS

3 tablespoons avocado oil

1 teaspoon ground coriander

1 teaspoon sweet paprika

1 teaspoon ground cumin

1 teaspoon freshly grated turmeric or ½ teaspoon ground turmeric

2 garlic cloves, crushed

1 cup stock of your choice (pages 280–81)

1 large (850 g) cauliflower, cut into florets

1 red onion, cut into wedges

8 apricots, halved and stones removed (or use 6 peaches, quartered, stones removed)

sea salt and freshly ground black pepper

ZUCCHINI COUSCOUS

4 zucchini, chopped

½ cup pistachio kernels

2 sprigs flat-leaf parsley, leaves and stems finely chopped

1 sprig mint, leaves picked, stems finely chopped

finely grated zest and juice of 1 lemon

Preheat the oven to 200°C (180°C fan-forced).

To make the roasted cauli and apricots, place all the ingredients in a bowl and toss to coat and combine well. Season well. Transfer the mixture to a heavy-based roasting tin. Roast for 40–45 minutes or until the stock has reduced by half and the cauliflower is tender and golden.

Meanwhile, to make the zucchini couscous, place the zucchini in a food processor and process until finely chopped, then transfer to a large bowl. Process the pistachios until finely chopped and add to the same bowl, along with the parsley, mint and lemon zest and juice. Season well, then toss together to combine. Set aside at room temperature, tossing occasionally, until you are ready to serve.

Divide the zucchini couscous evenly among serving bowls and spoon over the roasted cauliflower mixture to serve.

PUMPKIN and CHICKPEA CURRY

Chickpeas capture nitrogen from the atmosphere and 'fix' it in the soil, providing wonderful benefits to the land. This simple curry makes an ideal midweek meal and leftovers are great for lunch the next day. And if you can't get hold of a whole young coconut, don't worry – just add 1 cup of coconut water with the pumpkin and chickpeas.

SERVES 4

PREP: 25 MINUTES

COOKING: 20 MINUTES

3 tablespoons macadamia oil

2 garlic cloves, crushed

3 cm piece ginger, peeled and grated

2 cm piece turmeric, finely grated

1 tablespoon brown mustard seeds

2 sprigs curry leaves, leaves stripped

3 teaspoons ground cumin

2 teaspoons garam masala

650 g Kent pumpkin, cut into 3 cm pieces, skin left on

400 g can chickpeas, drained and rinsed

1 whole young coconut, top removed, water reserved and young flesh scraped from inside (optional)

sea salt and freshly ground black pepper

cooked brown rice, to serve (optional)

lime wedges, thinly sliced long red chilli, coriander leaves and coconut yoghurt, to serve

Heat the oil in a large, deep heavy-based frying pan over medium heat. Add the garlic, ginger, turmeric, mustard seeds, curry leaves, cumin and garam masala and cook, stirring constantly, for 2 minutes or until fragrant and the curry leaves are crisp.

Reduce the heat to medium–low and add the pumpkin, chickpeas and coconut water, stirring until well combined and coated in the spice mixture. Partially cover and simmer gently, stirring occasionally, for 12–15 minutes or until the pumpkin is tender and the coconut water has reduced by half.

Meanwhile, if you're using the young coconut, thinly shred the flesh.

Remove the pan from the heat, stir through the coconut flesh and season well. Divide the rice, if using, among serving bowls and top with the curry. Serve hot with the lime wedges, chilli, coriander and coconut yoghurt alongside.

FOUR AVOCADO SWEETS

Eating more avocado is great for our health and the health of the planet (as long as more forests aren't cleared to grow them!).

CHOC AVO MOUSSE WITH MIXED BERRIES

In a food processor, pulse 2 large avocados, 8 large pitted medjool dates and 3 tablespoons raw cacao powder for 20 seconds or until just combined. Add 3 tablespoons coconut cream and blend for 20 seconds or until smooth and fluffy, adding more coconut cream if needed to bring the mixture together. Divide among glasses and top with raspberries, blueberries and blackberries. Chill for at least 1 hour to set before serving.

AVOCADO LIME SLICE

Line the base and sides of an 18 cm × 8 cm loaf tin with baking paper. In a food processor, process together 1½ cups LSA (linseeds, sunflower seeds and almonds), 10 large pitted medjool dates and 3 tablespoons coconut oil until the mixture comes together in a ball. Press evenly over the base of the prepared tin. Blend 3 avocados, finely grated zest and juice of 3 limes and 3 tablespoons coconut oil for 1 minute or until smooth. Spread evenly over the date mixture. Chill overnight to set firm. Slice and serve with a spoonful of coconut yoghurt or chilled coconut cream.

PISTACHIO AND AVOCADO FUDGE

Line the base and sides of a 20 cm square cake tin with baking paper. In a food processor, process together 250 g LSA (linseeds, sunflower seeds and almonds), 12 large pitted medjool dates, 2 large avocados, 1 large overripe banana, ½ cup raw cacao powder and 1 teaspoon mixed spice until smooth, adding 1–2 tablespoons water if required. Press the mixture evenly over the base of the prepared tin, then press in ½ cup chopped, toasted pistachio kernels. Place in the freezer for 20 minutes or until firmed slightly, then cut into small squares. Return to the freezer for 20 minutes before serving.

PINE-AVO ICE POPS

Blend together 450 g chopped sweet pineapple, 1 avocado and 2–3 tablespoons milk of your choice until very smooth. Pour into six 180 ml ice-pop moulds and insert ice-pop sticks. Freeze overnight until firm.

CHILLI TOFU STIR-FRY

This stir-fry is great served with buckwheat, a nutrient-dense alternative to rice that can easily be cooked in big batches in advance, saving time and energy. To prepare it, cook raw buckwheat in a saucepan of boiling water over high heat for 25–30 minutes or until just tender, then drain. Rinse under cold running water, then drain well again and store in an airtight container in the fridge for up to 3 days or freeze portion-sized amounts in airtight containers for up to 2 months. When you're ready to serve, blanch the buckwheat in boiling salted water for 20–30 seconds, then drain.

SERVES 4

PREP: 20 MINUTES

COOKING: 15 MINUTES

2 tablespoons macadamia oil

150 g green beans, trimmed

2 long red chillies, finely chopped

3 cm piece ginger, peeled and finely chopped

300 g organic soft tofu, drained and broken into pieces

2 tablespoons tamari or soy sauce

4 green onions, diagonally cut into 4 cm lengths

¾ cup mint leaves

1 cup Thai basil leaves (or use regular basil)

sea salt and freshly ground black pepper

cooked buckwheat, to serve (optional)

Heat half the oil in a large wok over high heat. Add the beans and stir-fry for 4–5 minutes or until very crisp. Transfer to a bowl and set aside.

Heat the remaining oil in the wok over high heat. Add the chilli and ginger and stir-fry for 10 seconds, then add the tofu and stir-fry, breaking it into small chunks as you go, for 5 minutes or until golden.

Add the tamari and green onion and stir-fry for 30 seconds. Return the beans to the wok and toss until well combined. Remove the wok from the heat and toss through the mint and Thai basil leaves. Season well.

Serve the stir-fry over cooked buckwheat, if desired.

MIXED BEAN and MUSHROOM SHEPHERD'S PIE

One of the most nutritious and environmentally friendly comfort dishes you are likely to eat! You can add some nutritional yeast flakes (see page 261) to make it extra indulgent. If you find the cauliflower puree or bean mixture makes a little too much for your chosen baking dish, you can store them (separately) in airtight containers in the fridge for up to 3 days. The cauliflower puree can go in the freezer too (for up to 2 months), although the bean mixture won't freeze too well.

SERVES 4

PREP: 20 MINUTES, PLUS 5 MINUTES RESTING TIME

COOKING: 30 MINUTES

3 tablespoons extra virgin olive oil

1 onion, finely chopped

2 garlic cloves, crushed

1 carrot, finely chopped

1 celery stalk, finely chopped

150 g button mushrooms, sliced

3 tablespoons tomato paste

2 tablespoons tamari or soy sauce

400 g can red kidney beans, drained and rinsed

400 g can cannellini beans, drained and rinsed

1½ cups stock of your choice (pages 280–81)

1 (850 g) large cauliflower, florets separated, stem chopped

3–4 tablespoons milk of your choice

sea salt and freshly ground black pepper

Heat 2 tablespoons of the oil in a saucepan over medium heat. Add the onion, garlic, carrot and celery and cook, stirring occasionally, for 5 minutes or until softened. Add the mushroom and tomato paste and cook, stirring occasionally, for 5 minutes or until the mushrooms have softened.

Add the tamari or soy, beans and stock and stir until the mixture comes to a simmer. Continue to simmer, stirring occasionally, for 12–15 minutes or until the mixture has reduced and thickened.

Meanwhile, steam the cauliflower over boiling water for 15–18 minutes or until very tender. Cool for 5 minutes, then transfer to an upright blender. Add the milk and season well. Blend for 30–40 seconds or until completely smooth.

Preheat an oven grill to high. Transfer the bean mixture to a 2 litre ovenproof baking dish, then spoon over the cauliflower puree, making a pattern with a fork if you like. Drizzle the remaining oil evenly over the top and cook under the grill for 4–5 minutes or until the top is golden and crisp. Serve hot.

REFERENCES

37 A large chunk of our global emissions comes from making electricity (42 per cent including heat generation) . . .

International Energy Agency (IEA) 2016, *CO2 Emissions from Fuel Combustion* <https://www.iea.org/statistics/co2emissions/>.

37 Bangladesh is a country that understands the need to act fast to protect our environment.

World Bank Group 2013, *Turn Down The Heat: Climate Extremes, Regional Impacts and the Case for Resilience*, <http://www.worldbank.org/en/news/press-release/2013/06/19/warming-climate-to-hit-bangladesh-hard-with-sea-level-rise-more-floods-and-cyclones-world-bank-report-says>.

37 82 per cent of the costs of global warming are borne by poorer countries . . .

DARA & Climate Vulnerable Forum 2012, *Climate Vulnerability Monitor, A Guide to the Cold Calculus of a Hot Planet*, 2nd ed.

41 It is also estimated that each year, 240 million tonnes of carbon dioxide enters the atmosphere from kerosene . . .

Jacobson, A. et al. 2013, *Black Carbon and Kerosene Lighting, An Opportunity for Rapid Action on Climate Change and Clean Energy for Development*, Brookings Institute.

47 That's $10 million a minute.

Coady, D et al. 2015, How Large Are Global Energy Subsidies?, International Monetary Fund (IMF).

47 Research from the Centre for the Understanding of Sustainable . . .

Stone L 2018 *From stranded workers to enabled workers—Lessons for a successful low carbon economy*, Centre for the Understanding of Sustainable Prosperity (CUSP), <https://www.cusp.ac.uk/themes/aetw/lowcarbon-agulhas-report>

48 The only downside was that managers and executives (comprising 3.2 per cent of coal workers) would make less.

Louie, EP & Pearce JM 2016, 'Retraining investment for U.S. transition from coal to solar photovoltaic employment', *Energy Economics*, vol. 57, pp. 295–302.

49 Today's amount of renewable energy is 40 times larger than the IEA predicted it would be in 2003.

Intergovernmental Panel on Climate Change (IPCC) 2018, *Sixth Assessment Report (AR6): Working Group 111.*

59 From 1990 to 2010, 95 per cent of the wealth created remained with just 40 per cent of the population.

Hickel, J 2017, *The Divide: A Brief Guide to Global Inequality and its Solutions*, William Heinemann. Also Woodward D 2015, 'Incrementum ad absurdum: Global Growth, Inequality and Poverty Eradication in a Carbon-Constrained World', *World Social and Economic Review*, vol 4.

59 CEO salaries have increased by almost 1000 per cent . . .

Mishel L & Davis A 2015, 'CEO Pay Has Grown 90 Times Faster than Typical Worker Pay Since 1978', *Economic Policy Institute.*

61 Research in recent decades has shown that inequality is incredibly significant.

Wilkinson RG & Picket K 2009, *The Spirit Level: Why More Equal Societies Almost Always Do Better*, Allen Lane.

66 Our global GDP (the total value of goods and services in all the world's economies) is around US$80 trillion.

World Bank Data Bank <http://databank.worldbank.org/data/download/GDP.pdf>.

67 A sustainable level of resource use is around 50 billion metric tons per year . . .

Dittrich M et al. 2012, 'Green Economies Around the World? Implications of Resource Use for Development and the Environment', *Sustainable Europe Research Institute* (SERI).

73 Emissions from road vehicles are one of the largest contributors to our sick planet . . .

US Environmental Protection Agency 2018, *Fast Facts U.S. Transportation Sector Greenhouse Gas Emissions 1990–2016* <https://nepis.epa.gov/Exe/ZyPDF.cgi?Dockey=P100U[illegible]>

73 By 2040, an extra billion cars are predicted to join the 1.2 billion already on the planet.

Bernstein Research House, *The Number of Cars Worldwide is Set to Double by 2040* <https://www.weforum.org/agenda/2016/04/the-number-of-cars-worldwide-is-set-to-double-by-2040>.

76 In just 10 years of operation Uber is creating more bookings in the US than the country's entire taxi industry.

Seba, T 2014, *Clean Disruption of Energy and Transportation: How Silicon Valley Will Make Oil, Nuclear, Natural Gas, Coal, Electric Utilities and Conventional Cars Obsolete by 2030.*

76 In fact, 20 per cent of all miles travelled in San Francisco are now in either Uber or Lyft vehicles.

Seba, T 2014, *Clean Disruption of Energy and Transportation: How Silicon Valley Will Make Oil, Nuclear, Natural Gas, Coal, Electric Utilities and Conventional Cars Obsolete by 2030.*

87 Research has found that if global cycling rates can . . .

Mason, J, Fulton L & McDonald Z 2015, *A Global High Shift Cycling Scenarion: The potential for DramaticallyIncreasing Bicycle and E-bike Use in Cities Around the World, with Estimated Energy, CO2 and Cost Impacts*, Institute for Transportation & Development Policy UC Davis.

88 Statistics show that if everyone in the US stopped . . .

'Green is my thing' website 2015 <https://greenismything.com/2015/09/16/what-if-everyone-in-the-u-s-stopped-driving-for-one-day/>.

89 An economy return flight from Sydney to London uses . . .

Potter, A 2017, 'Should you buy carbon offsets?' *CHOICE* <https://www.choice.com.au/travel/on-holidays/airlines/articles/should-you-buy-carbon-offsets-for-flights>.

89 Less than 10 per cent of Australian travellers choose . . .
Smith, B 2018, 'Is it worth paying for carbon offsets?' *ABC News Science* <https://www.abc.net.au/news/science/2018-04-11/carbon-offsets-worth-buying-air-travel-tourism-emissions/9638466>.

90 These companies were well aware of the damage their products were doing to the planet . . .
Franta, B 2018, 'Shell and Exxon's secret 1980s climate change warnings', *The Guardian* <https://www.theguardian.com/environment/climate-consensus-97-per-cent/2018/sep/19/shell-and-exxons-secret-1980s-climate-change-warnings>.

90 The Libertarian's platform called for the abolition of all government healthcare programs . . .
Mayer, J 2016, *Dark Money*, Penguin Random House.

91 Agnotology is the study of how ignorance is spread . . .
Bedford, D 2010, 'Agnotology as a Teaching Tool: Learning Climate Science by Studying Misinformation', *Journal of Geography*, vol. 109, no. 4, pp. 159–165, DOI:10.1080/00221341.2010.498121.

93 There is a 97 per cent consensus among climate scientists that global warming is real and that we are causing it.
Cook, J et al. 2016, 'Consensus on consensus: a synthesis of consensus estimates on human-caused global warming', *IOP science*, Environmental Research Letters, vol. 11, no. 4.

97 Even if all emissions stopped right now, global warming would still continue for centuries . . .
NASA 2018, 'Responding to Climate Change' <https://climate.nasa.gov/solutions/adaptation-mitigation/>.

102 Land clearing and deforestation for agriculture makes up 15–18 per cent of our food system's greenhouse gas . . .
World Wildlife Fund <https://www.worldwildlife.org/threats/deforestation>.

102 The livestock sector contributes 14.5 per cent of total global emissions.
Rojas-Downing, M et al. 2017, 'Climate change and livestock: Impacts, adaptation, and mitigation', *Climate Risk Management*, vol. 16, pp 145–63, DOI:10. 1016/j.crm.2017.02.001.

102 Since 1850, the clearing of land plus constant ploughing or tilling of the soil has released 155 billion tonnes . . .
Toensmeier, E 2016, *The Carbon Farming Solution: A Global Toolkit of Perennial Crops and Regenerative Agriculture Practices for Climate Change Mitigation and Food Security*, Chelsea Green Publishing Co.

102 A combination of studies suggest that somewhere between 5 and 7 billion animals . . .
Fischer, B & Lamey, A 2018, 'Field Deaths in Plant Agriculture', *Journal of Agricultural and Environmental Ethics*, vol. 31, pp. 409–28, DOI:10.1007?s10806-018-9733-8.

106 A spoonful of healthy soil can contain . . .
Ingham, ER, 'Soil Bacteria', *USDA Natural Resources Conservation Service* <https://www.nrcs.usda.gov/wps/portal/nrcs/detailfull/soils/health/biology/?cid=nrcs142p2_053862>.

113 70–80 per cent of Amazon deforestation is for cattle . . .
Yale School of Forestry & Environmental Studies, Global Forest Atlas, *Cattle Ranching in the Amazon Region* <https://globalforestatlas.yale.edu/amazon/land-use/cattle-ranching>.

113 Methane (from all sources) makes up 16 per cent of total greenhouse gas emissions, with methane from livestock making up around 44 per cent of that figure:.
New Zealand's Environmental Reporting Series: Environmental indicators <http://archive.stats.govt.nz/browse_for_stats/environment/environmental-reporting-series/environmental-indicators/Home/Atmosphere-and-climate/global-greenhouse-gases.aspx >Food and Agriculture Organization of the United Nations <http://www.fao.org/news/story/en/item/197623/icode/>.

114 The issue of methane and feedlots is complicated . . .
Capper, JL 2012, 'Is The Grass Always Greener? Comparing the Environmental Impact of Conventional, Natural and Grass-Fed Beef Production Systems', *Animals (Basel)*, vol. 2, no.2, pp. 127–43.

116 A Colombian study showed intensive silvopasture can sequester 8.8 tonnes. . .
Cuartas Cardona CA et al. 2014 'Contribution of intensive silvopastoral systems to animal performance and to adaptation and mitigation of Climate Change', *Revista Colombiana de Ciencias Pecuarias*, vol. 27, pp. 76–94.

118 In the US the proportion is 40 per cent.
Gunders, D 2012, 'How America is losing up to 40 per cent of its food from farm to fork to landfill.' Natural Resources Defense Council (NRDC).

118 Worldwide we waste around 30 per cent of our food (through production, transport and consumption.
Food and Agriculture Organization of the United Nations, *Save Food Global Initiative on Food Loss and Waste Reduction* <http://www.fao.org/save-food/resources/keyfindings/en/>

123 We have been extracting life from them for centuries, to the point where 90 per cent of the world's fish stocks are over-fished.
Food and Agriculture Organization of the United Nations, *The State of the World's Fisheries and Aquaculture* < http://www.fao.org/3/a-i5555e.pdf>

130 Even as deforestation persists, the regrowth of tropical forests can sequester . . .
Hawken, P 2017, *Project Drawdown: The Most Comprehensive Plan Ever Proposed to Reverse Global Warming*, Penguin Random House.

131 They make up just 3 per cent of the Earth's landmass but are the second biggest storer of carbon after oceans.
Hawken, P 2017, *Project Drawdown: The Most Comprehensive Plan Ever Proposed to Reverse Global Warming*, Penguin Random House.

131 Composting food scraps can really improve the carbon-sequestering ability of soil while also preventing more greenhouse gases from entering the atmosphere.

Favoino, E & Hogg, D 2008, 'The potential role of compost in reducing greenhouse gases', *Waste Management Research,* vol 26, no 61.

138 You can even turn your backyard into a carbon-sequestering food hub . . .

The Climate Reality Project 2017, *Right Under Your Feet: Soil Health and the Climate Crisis* <https://www.climaterealityproject.org/content/thank-you-downloading-soil-health-and-climate-crisis>.

153 They believe we have to scale down global material consumption by at least 20 per cent . . .

Intergovernmental Panel on Climate Change (IPPC) 2018, 'Special Report on Global Warming of 1.5°C (SR15)' <http://www.ipcc.ch/report/sr15/>

172 In Bangladesh, the fertility rate (average births per woman) has dropped from six in the 1980s to two . . .

Hawken, P. 2017, *Project Drawdown: The Most Comprehensive Plan Ever Proposed to Reverse Global Warming,* Penguin Random House.

172 According to a recent Yale study, 70 per cent of people in the US now believe protecting the environment . . .

Marlon, J et al. 2018, *Yale Climate Opinion Maps,* Yale Program on Climate Change <climatecommunication.yale.edu/visualizations-data/ycom-us-2018/?est=happening&type=value&geo=county>

172 Scientists have identified that just 10 rivers in the . . .

Schmidt, C, Krauth, T & Wagner, S 2017, 'Export of Plastic Debris by Rivers into the Sea', *Environmental Science & Technology,* vol. 14, no. 21.

174 A couple of years ago, my wife, Zoe, read . . .

Kondo, M 2014, *The Life-Changing Magic of Tidying Up,* Ten Speed Press.

191 Studies show that the important choices we make in our homes (in regards to electricity use and the foods we eat) coupled with our choices . . .

Jones, CM, Wheeler SM & Kammen, DM 2018, 'Carbon Footprint Planning: Quantifying Local and State Mitigation Opportunities for 700 California Cities', *Urban Planning,* vol. 3, No. 2.

191 Studies also show that when a home installs solar panels on their roof, the odds of a neighbour installing . . .

Bollinger, B. & Gillingham, K 2012, 'Peer Effects in the Diffusion of Solar Photovoltaic Panels', *Marketing Science,* vol. 31, no. 6, pp.873–1025.

193 No campaign failed once it achieved sustained and active participation of just 3.5 per cent of the population.

Chenoweth, E & Stephan, MJ 2012, *Why Civil Resistance Works: The Strategic Logic of Nonviolent Conflict,* Columbia University Press.

196 Interestingly, 81 per cent of people in now Britain believe that the government's prime objective should be . . .

BBC 2006, 'The Happiness Formula', Question 14 on opinion poll conducted by GfK NOP Ltd.

199 Between 70 and 80 per cent of the food in the world comes from smallholders and the rest from industrial agriculture.

Food and Agriculture Organization of the United Nations, *The State of Food and Agriculture,* 2014 < http://www.fao.org/3/a-i4040e.pdf>

200 Globally, more than $600 billion is spent on advertising each year.

eMarketer 2014, 'Advertisers Will Spend Nearly $600 Billion Worldwide in 2015'.

200 Americans are now purchasing double what they did in 1950, and the average child in the US watches 40,000 commercials a year.

Committee on Communications 2006, 'Children, Adolescents, and Advertising', *American Academy Of Pediatrics,* vol. 118, no. 6, pp. 2563–69.

202 A recent study showed that around 66 per cent of the Make America Great Again hashtags (#MAGA) during the 2016 US election were coming from bots.

Schwab, K 2017, 'These Students Built The Anti-Bot Algorithm Twitter Desperately Needs', *Fast Company.*

202 In the middle was what they called an 'exhausted majority' who were fed up with the polarisation.

More in Common 2018, *Hidden Tribes: A Study of America's Polarized Landscape* <https://www.moreincommon.com/hidden-tribes/>.

206 In the US, a whopping $3.3 billion per year is spent lobbying politicians . . .

Statista 2018, *Total lobbying spending in the United States from 1998 to 2017.*

206 For every $1 that public interest groups spend lobbying, corporations spend $34.

Drutman, L 2015, 'How Corporate Lobbyists Conquered American Democracy', *The Atlantic.*

206 In contrast, the war chest accumulated by the Koch brothers and their friends for the Republican party in the 2016 election was around US$889 million.

Confessore, N 2015, 'Koch Brothers' Budget of $889 Million for 2016 is on Par With Both Parties' Spending' *The New York Times* < https://www.nytimes.com/2015/01/27/us/politics/kochs-plan-to-spend-900-million-on-2016-campaign.html>.

207 When the banks needed to be bailed out after the crash in 2008, at least $8 trillion was found (some estimate $16 trillion, even $29 trillion) to tidy up the mess because the banks were 'too big to fail'.

Collins, M 2015, 'The Big Bank Bailout', *Forbes.*

CONVERSION CHART

Measuring cups and spoons may vary slightly from one country to another, but the difference is generally not enough to affect a recipe. All cup and spoon measures are level.

One Australian metric measuring cup holds 250 ml (8 fl oz), one Australian tablespoon holds 20 ml (4 teaspoons) and one Australian metric teaspoon holds 5 ml. North America, New Zealand and the UK use a 15 ml (3 teaspoon) tablespoon.

LENGTH

Metric	Imperial
3 mm	⅛ inch
6 mm	¼ inch
1 cm	½ inch
2.5 cm	1 inch
5 cm	2 inches
18 cm	7 inches
20 cm	8 inches
23 cm	9 inches
25 cm	10 inches
30 cm	12 inches

LIQUID MEASURES

One American pint	One Imperial pint
500 ml (16 fl oz)	600 ml (20 fl oz)

Cup	Metric	Imperial
⅛ cup	30 ml	1 fl oz
¼ cup	60 ml	2 fl oz
1/3 cup	80 ml	2½ fl oz
½ cup	125 ml	4 fl oz
⅔ cup	160 ml	5 fl oz
¾ cup	180 ml	6 fl oz
1 cup	250 ml	8 fl oz
2 cups	500 ml	16 fl oz
2¼ cups	560 ml	20 fl oz
4 cups	1 litre	32 fl oz

DRY MEASURES

The most accurate way to measure dry ingredients is to weigh them. However, if using a cup, add the ingredient loosely to the cup and level with a knife; don't compact the ingredient unless the recipe requests 'firmly packed'.

Metric	Imperial
15 g	½ oz
30 g	1 oz
60 g	2 oz
125 g	4 oz (¼ lb)
185 g	6 oz
250 g	8 oz (½ lb)
375 g	12 oz (¾ lb)
500 g	16 oz (1 lb)
1 kg	32 oz (2 lb)

OVEN TEMPERATURES

Celsius	Fahrenheit
100°C	200°F
120°C	250°F
150°C	300°F
160°C	325°F
180°C	350°F
200°C	400°F
220°C	425°F

Celsius	Gas mark
110°C	¼
130°C	½
140°C	1
150°C	2
170°C	3
180°C	4
190°C	5
200°C	6
220°C	7
230°C	8
240°C	9
250°C	10

INDEX

THANK YOU TO:

My supportive family, particularly Zoë and Velvet, Jeff, Jill and Miriam.

The tireless Pan Mac team: Ingrid Ohlsson, Ariane Durkin, Naomi van Groll, Sally Devenish, Belinda Huang, Georgia Webb, Rebecca Lay and Charlotte Ree.

And lovely freelancers: Arielle Gamble, Miriam Cannell, Trisha Garner, Tracey Pattison, Rebecca Hamilton, Rachel Carter, Susanne Geppert, Alice Oehr and Luke Bubb.

All the *2040* film cast and crew, but especially Anna Kaplan, Billy Wychgel and Jenna Dawkins, who also helped with the book.

All of the scientists and experts who gave their time, responded to my emails and corrected my mistakes, especially Paul Hawken, Eric Toensmeier, Tony Seba and Prof Lesley Hughes.

A special mention to Ian Darling for enabling so much of the *2040* journey to come to life.

First published 2019 in Macmillan
by Pan Macmillan Australia Pty Limited
1 Market Street, Sydney, New South Wales
Australia 2000

A CIP catalogue record for this book
is available from the National Library of Australia:
http://catalogue.nla.gov.au

Design by Arielle Gamble and Trisha Garner
Design assistance by Susanne Geppert
Illustrations by Arielle Gamble
Photography by Cath Muscat and Rob Palmer
Recipe development by Tracey Pattison
Prop and food styling by Michelle Noerianto
Food preparation by Kerrie Ray and Sarah Mayoh
Editing by Miriam Cannell and Rachel Carter
Index by Helena Holmgren

Colour + reproduction by Splitting Image Colour Studio
Printed in China by 1010 Printing International Limited